JUPITER—
THE STAR THAT FAILED

WESTMINSTER PRESS BOOKS
BY
JOEL N. SHURKIN

Jupiter—The Star That Failed

Update—Report on the Planet Earth

JUPITER—
THE STAR THAT FAILED

by
Joel N. Shurkin

THE WESTMINSTER PRESS
Philadelphia

All photographs are by courtesy of the National Aeronautics and Space Administration.

BOOK DESIGN BY DOROTHY ALDEN SMITH

First edition

Published by The Westminster Press®
Philadelphia, Pennsylvania

PRINTED IN THE UNITED STATES OF AMERICA

9 8 7 6 5 4 3 2 1

Library of Congress Cataloging in Publication Data

Shurkin, Joel N 1938
Jupiter, the star that failed .

Includes index.
SUMMARY: Discusses our solar system's largest planet, including Galileo's observations of it with the first telescope and revealing data supplied by the recent Pioneer X space probe.
1. Jupiter (Planet) — Juvenile literature.
[1. Jupiter (Planet) 2. Pioneer Project. 3. Space flight to Jupiter] I. Title.
QB661.S58 523.4′5 78–10885
ISBN 0-664-32642-0

To Fannie Lampf, with love

Contents

1
The Birth of a Giant

To understand the strange giant, you have to know how all things were born.

We do not know how or exactly when time or the universe began, but we have never been slow about guessing. We haven't been slow about trying to explain the stars, the moon, and the sun either. From the days of mythology, human beings have devised uncounted numbers of scenarios about how the universe works and how it was born, and we're still at it.

Once it was thought that humankind was at the heart of all things. The Earth was flat, we believed, and it was the center of the universe. The sky circled overhead, the stars glowed, little points on the ceiling of creation.

We have tried to explain the creation and the creator in various ways. The Norse, for instance, believed that at first the universe was empty. "Earth was not found, nor Heaven above, a yawning-gap there was, but grass no where." Grass was important to the story because the first two creatures to arrive on the scene were a giant, Ymer, and a cow, Audhumla. Everything on earth came from those two characters.

In explaining the creator, the nineteenth-century Japanese writer Hirata Atsutane assured us in *Summary of the Ancient Way* that "the gods who formed all the lands of the world at the Creation . . . were without exception born in Japan. Japan is thus the homeland of the gods, and that is why we call it the Land of the Gods." He did not explain who created Japan.

Some people thought that the creation had not been done very well. Alfonso the Wise of the thirteenth century wrote: "Had I been present at the creation, I would have given some useful hints for the better ordering of the universe."

Perhaps the best-known description comes from the Old Testament:

> In the beginning God created the heaven and the earth. And the earth was without form, and void; and darkness was upon the face of the deep. And the Spirit of God moved upon the face of the waters. And God said, Let there be light; and there was light.

Remarkably, that description from Genesis 1:1 sounds very much like the current scientific scenario of the creation, a resemblance that has been noticed by a number of astronomers.

Scientific scenarios are called "models," and the current standard model for the creation of the universe is called "the big bang." It goes like this:

In the beginning there was the explosion. The explosion was all over space. Everything was exploding. There was no center to the blast. Tiny particles, the kinds of things that now make up atoms and molecules, were blown around madly. They could not form atoms then because the unimaginable power of the big bang produced temperatures so high (100,000,000° Celsius) that they could never get together long enough to unite. It was the explosion of creation. It was the source of all things.

There were no elements or gases in the fraction of a second after the bang, only those pieces of atoms flying around.

And there was light. In fact the light was everywhere. The whole universe shone with a light that is beyond our imagination. Scientists now know that light is also made up of particles called photons. They have no electric charge, unlike most of the other particles moving around then, and they have no mass (which is like saying they are weightless), but they contain enormous amounts of energy.

As quickly as all these particles were formed, they self-destructed, only to be replaced by new ones from the immense energy that was everywhere. Space was thick with them, millions and millions of times thicker than water, an unthinkably hot, bright soup.

Occasionally a heavy particle like a proton or a neutron roared around, but mostly the universe at that second was made up of electrons, their mysterious alter egos, the positrons, ghostly little neutrinos, and photons, rushing insanely from each other, annihilating with energy and being reborn. Astronomer Robert Jastrow has written that the giant explosion and the light is as good a definition of God as scientists will ever find. He thinks the similarities between the big bang theory and Genesis are no coincidence.

A fraction of a second after that cosmic blast, the universal temperature began to drop to around 30 billion degrees. By creation +1 second it was down to 10 billion, and 14 seconds later to 3 billion. After about 3 minutes it was down to 1 billion degrees, and some order was coming out of the chaos. Now it was "cool" enough for some of the particles to get together and form atoms, beginning with the simplest, hydrogen (with a nucleus of one proton and a neutron in its first, or heavy, form, plus an electron), and helium (two protons, two neutrons, and two electrons). The two elements would soon form vast clouds that spread out all over creation, clouds that would eventually condense into galaxies.

That explosion was about 20 billion years ago, give or take a couple of billion years. The process that formed the galaxies themselves took billions of years. Think of gigantic clouds of gas, each cloud made up of particles of matter, usually molecules of hydrogen and helium. Each particle has a gravitational pull. Each is attracted to another particle as it floats around, which in turn attracts still another. Soon the cloud is beginning to form clumps. It is becoming an enormous field of smaller, condensing clouds, each one destined to be a star, a planet, or a moon.

From those clouds came our world, and the giant we are about to study.

What the cloud will turn out to be depends on how much material is contained in it. The more material, measured in terms of mass, (or weight, to simplify things), the more likely it is that the cloud will become a star.

As the particles gravitate toward one another, the unborn star gets thicker in the middle, forming a center that draws in even more particles. As the particles get closer and closer together, they begin to rub against each other and get hotter. The more particles are drawn in, the hotter the center gets. Finally, there is enough matter pulled together to bring the temperatures to a magic point—when the combination of heat and pressure forces the hydrogen atoms to fuse together. The result is an atom of helium and enormous amounts of energy in the form of heat and light. The cloud bursts into brilliant flame, and a star is born.

The process is almost self-destructive. The intensity of the fiery outburst puts out the blaze, and the star must contract and expand again before it finally stabilizes into a bright shining star. Our sun went through just such a period.

In the neighborhood of our sun, as it was in the process of stabilizing 4.5 billion years ago, were several other, smaller clouds. They too were round, swirling outward from a center. They were also contracting. One of them, larger than the rest and certainly a giant in its own right, was beginning to go through the same birth processes as the sun, but this one could not quite make it. Large as it was, the giant was not big enough to touch off fusion ignition. In fact it had only 1/70th (1.4 percent) of the required mass. But its nature would be the result of that failure.

This round cloud was still gigantic; it was big enough to get very hot and glow a bright red. Temperatures hit 50,000° C at the center and 1,500° C at the outer edge as the cloud condensed into a huge ball.

What happened next is a matter of some conjecture. There are two models, one depending on that ball having no rocky core, the other requiring a core. We may never know which one is right.

What does seem clear to scientists is that the ball went through timeless periods of contraction and expansion, getting hotter and then cooler.

In the first contraction, the size of the ball of gas shrank from 966 million kilometers to about 644 million km in less than a million years. For events in astronomy, that is very fast work.

But events moved even faster. The ball continued to shrink, dropping down to about its present diameter, around 161,000 km. At this point it probably had a composition much like that of the sun. Of its atoms, 93 percent were hydrogen, 6 percent were helium, and the rest, 1 percent, were the other elements that evolved out of the hydrogen and helium.

The ball was very cold, perhaps only 100 degrees warmer than absolute zero, the point at which all molecules freeze into stillness. But finally, the contraction caused another flare-up. Temperatures shot up to around 2,210° C and the ball greatly expanded. The cycle continued.

The expansion led to a dramatic event—a collapse that may have lasted only three months in our time. In astronomy almost nothing is ever measured in times that short. The ball stabilized, after that quick shrink, at about 644,-000 km. It was still producing great amounts of heat and was glowing red hot. Temperatures in the center were probably around 50,000° C. It was a dull red giant in the primeval skies.

This hot period lasted for about 100,000 years. Contraction and cooling slowed down. The process, however, is still going on today.

If the planet has a rocky core, the model is a bit different in the beginning. Dust grains in the collapsing cloud would have been caught by lighter particles of gas and attracted to each other by gravity. They formed a center in the cloud and eventually grew into a ball of rock that was about 30 times the size of today's Earth. Lighter elements in the cloud would have fallen onto the core. The heat and contraction period would be much the same as in that first, liquid, model.

While all this was going on, in either model, smaller clouds of gas were circling the large cloud, trapped by gravity in eternal orbit. They were also going through contraction and condensation. There were more than a dozen of those clouds, destined to be the moons of the planet being formed.

The largest clouds contained a great many heavy elements such as those which formed the earth and moon: silicon, iron, magnesium, and oxygen. There was also a great deal of water ice.

The smaller, outer satellites consisted of lighter, more sunlike material.

The dozen or so moons (scientists are still discovering moons) were now circling the largest planet in the solar system. A breeze of tiny particles from the sun, called the *solar wind,* blew away all the loose material that had not condensed into the planet or its moons, cleaning the area like a fan cleaning away dust. The stage was set.

So the planet Jupiter was created because it was a star that failed.

2
Marduk

The first humans, whoever they were, must have looked up at the night sky and wondered. They saw the stars sprawled across the sky. They saw day follow night and another new day born. The sun rose and set, the moon came and went in many shapes. But until relatively recently in human history the only tool that people had to learn about the heavens was the naked eye. It was not a very good tool, but sometimes it worked quite well.

Each person watched long enough to derive many theories about what appeared in the skies. Most of them sound pretty silly to us now, but they are, actually, logical conclusions drawn from simple observations. Some of the conclusions were remarkably accurate.

Not until the invention of the telescope could real evidence be gathered to prove or disprove the conclusions drawn from the naked eye. Radiotelescopes and other delicate electronic gear vastly increased our knowledge. But as in the case of Jupiter it was the use of rockets and unmanned space probes that really began the exploration of Earth's neighboring worlds in space.

Scientists like to say that time and space are related. Man's measurements of space are probably related to time as well. The first astronomical observations were probably made when early people began to tell time by watching heavenly bodies. Historian Will Durant writes that the English words for "measure," "month," and "man" probably stemmed from the word for "moon." The changes in the moon were probably the very first measure of time. It was easier to describe time in relation to moons ("many moons ago") than by years. Later, the moon did provide the measure for months and years, and even today many calendars, like the Jewish calendar, use the moon as the basis for a month. A lunar calendar is not so accurate as the solar calendar, as men soon learned.

The stars quickly developed a mystical meaning for prehistoric people. Astrology, which claims to be able to find an influence on people exerted by the stars, became, in a way the first "science." If you pick up any daily newspaper, you will find there are still a lot of believers in astrology. They are totally unimpressed with evidence that astrology has no basis in fact.

Sooner or later the ability of humankind to count and to perform simple mathematical functions added to the ability to keep track of what was going on in the sky.

The Egyptians of 2000 B.C. could predict the floods of the Nile, a vital event, because the floods produced the crops that fed Egypt. The Egyptians could tell when to plant the crops, when to expect the watering floods, and when to start worrying if the floods didn't appear. They also built their temples to face the exact place on the horizon where the star Sirius would appear at the summer solstice, the day summer begins, June 21. They may have even been able to trace the movements of the planets, which could only be discerned by watching them over a long period of time. So they were probably the first people to know that those lights in the sky were not all the same.

The Egyptians eventually wound up with thousands of years of astronomical data.

The most famous of the ancient astronomers were the Babylonians. They also based their science on astrology—

Jupiter crescent. A view of the planet "Marduk" never seen from Earth

and actually they only incidentally wound up with astronomy—but they too found that some of the stars they saw were different from most of the others and gave names to five of the planets. Mercury was Nabu, Venus was called Ishtar, Mars was Nergal, Saturn was called Ninib, and Jupiter, whose path across the skies they plotted quite accurately, was known as Marduk. They could tell in advance when Marduk would appear in the skies and in which part of the sky it would first rise. They also found the astronomical seasons, including the equinoxes and the solstices, and divided the sky into ten signs of the zodiac, "a circle of living things."

But the Babylonians too were interested mostly in the magic of the sky and the influence of the stars on people's lives. Astrologers became the richest men in town, and even though they were outright quacks, there were enough real sky watchers among them to invent some of the astronomical techniques we still use today.

For instance, they were the first to divide the sky into 360 degrees and divide each degree into minutes and hours, a system still used by astronomers.

It is hardly surprising that many of the ancients worshiped some of the heavenly bodies as gods, or named them after the gods.

The Greeks were so good at astronomy that they provided many cultures with their science and with the religions that derived from it. In India, for instance, the earliest astronomical reference book was the *Siddhantas* (425 B.C.), which was really just a Hindu version of Greek astronomy. On the other hand, the Indians were good enough to develop a system of plane geometry from their astronomical observations, and they also correctly guessed that a spinning Earth produced night and day. They discovered, too, or at least described, gravity.

All this was done with naked-eye observations, because it was not until the early seventeenth century that telescopes were invented.

Just before that invention, the scientific world was ripe for revolution. It often happens that way. Intense turmoil builds for years. Suddenly there is a breakthrough, as if someone punches a hole in a dam to let loose a torrent.

Through the Middle Ages science and especially astronomy were in the grip of the church. The church would not let go of old doctrines, under the belief that they were divinely inspired. To challenge an old doctrine—for instance, the doctrine that holds the Earth as the center of the universe—was considered heresy.

The idea of the universe revolving around the Earth was not directly challenged until the middle of the sixteenth century, when a Pole, Mikolaj Kopernik (better known as Copernicus) published a book which stated that the Earth really revolved around the sun. There was a predictable furor, but Copernicus did not live long enough to suffer from it. Other men would.

The unaided eye was not a bad instrument for poking into the mysteries of the universe. But it was not very good at solving them. You could see that there are stars and planets, the planets behaved differently from stars (they did not blink and they moved faster). You could see some of the irregularities in the surface of the moon, but you could not begin to understand what they were. You could spot meteors and comets, but you could not know what they were either. And you couldn't look at the sun at all.

Clearly, the eye needed assistance.

Believe it or not, no one knows for sure who invented the telescope. Most give credit to Hans Lippershey, a lens grinder from Walcheren, Holland. The lenses that Lippershey worked on were the kind used for eyeglasses, which had been invented sometime in the thirteenth century.

Little is known about Lippershey except for one fact: on October 2, 1608, he produced a telescope for the Dutch government. They asked him for a two-eyed version and he came up with three of them—the first binoculars. A year later you could buy telescopes in Paris. According to English astronomer Patrick Moore, the telescope was designed for looking at things on earth close-up, but one of the sales brochures advised that through this invention you might be able to glimpse stars that would normally be invisible.

Astronomers had no trouble making that connection, and one of them, an Italian named Galileo Galilei, went to work immediately. He built his own telescope in 1609. The government of his native Venice was thrilled at the possibilities, and Galileo was given a teaching job at the University of Padua. He eventually made about one hundred telescopes. You can see some of them today in museums.

Each of his telescopes had two lenses. One was convex (bulged outward) and the other, the eyepiece, was concave (curved inward). These were not very good, magnifying only around 30 times, and the field of vision was very small.

As crude as Galileo's telescopes were, they revolutionized astronomy and got Galileo into a heap of trouble.

By 1610, he was watching the stars almost every night. He quickly discovered that the planets were round objects that could be seen enlarged in the telescope, while the stars looked exactly the same as to the naked eye. (We now know that the stars are so far away they cannot be enlarged visually.) But he also noted that many stars that were once invisible could now be seen through his lenses. And there were zillions of those stars.

"Beyond the stars of the sixth magnitude, you will behold through the telescope a host of other stars, which escape the unassisted sight, so numerous as to be almost beyond belief," Galileo wrote. He found 500 new stars in one tiny section of the constellation Orion alone.

He spotted the craters on the moon and described some of the lunar surfaces.

He even found one planet with other moons. It was our giant, Jupiter.

Here is Galileo's description of the discovery:

> On the 7th day of January in the present year, 1610, in the first hour of the following night, when I was

> viewing the constellations in the heavens through a telescope, the planet Jupiter presented itself to my view, and as I had prepared for myself a very excellent instrument, I noticed a circumstance which I had never been able to notice before, namely, that three little stars, small but very bright, were near the planet; and although I believed them to belong to the number of fixed stars, yet they made me somewhat wonder, because they seemed to be arranged exactly in a straight line, parallel to the ecliptic, and to be brighter than the rest of the stars equal to them in [apparent] magnitude. . . .
>
> On the east side there were two stars, and a single one toward the west. The star which was furthest toward the east, and the western star, appeared rather larger than the third.
>
> I scarcely troubled at all about the distance between them and Jupiter, for, as I have already said, at first I believed them to be fixed stars; but when on 8 January, led by some fatality, I turned again to look at the same part of the heavens, I found a very different state of things, for there were three little stars all west of Jupiter, and nearer together than on the previous night, and they were separated from one another by equal intervals. . . .

He kept looking night after night and found that on every night the combination of little stars changed in relation to the planet.

> I therefore concluded, and decided unhesitatingly, that there are three stars in the heaven moving about Jupiter, as Venus and Mercury round the sun; which at length was established as clear as daylight by numerous other subsequent observations. These observations also established that there are not only three, but four, erratic sidereal bodies performing their revolutions around Jupiter. . . .

He named the four moons he discovered after the Medicis, the leading political and intellectual lights of seventeenth-century Italy, but they were later renamed for mythological creatures, Io, Europa, Ganymede, and Callisto.

Galileo later discovered something odd about Saturn (we know now it was the rings), and found a sensible way to study the sun through a telescope, by reflecting the image on a piece of paper. That led to the discovery of sunspots.

He turned his telescope then toward Venus and, by studying how the sun shone on only parts of the disk at any one time, he came up with observable proof that Copernicus was right—the planets orbit the sun. He could also demonstrate that the planets had moons.

All of Galileo's discoveries made sense only if Copernicus' view of the universe was correct, and Galileo made no secret of that fact. He published the results far and wide, and the church, fighting a last-ditch effort to save the old from the new, fought back until Galileo was forced to recant. He was kept in virtual house arrest the remaining years of his life and died in solitude and blindness on January 8, 1642, thirty-two years to the day after he discovered the moons of Jupiter.

Galileo won the argument and the church has never been able to completely justify its action, but the dam had been broken. People could now actually look at the stars and the worlds of space.

There were more discoveries to follow, such as Isaac Newton's reflector telescope, which greatly aided in the observation of the planets.

Eventually, astronomers learned that they did not have to rely on light alone to watch the stars. They could use other forms of radiation, like radio waves, to draw pictures of the objects that twinkled and shone in the sky.

In 1870 a Scottish mathematician, James Clerk Maxwell, discovered that what we think of as electricity and magnetism are really one and the same thing. Called "electromagnetism," it meant that there was a range of radiation in nature far longer (infrared) and shorter (ultraviolet) than our eyes can make out. His formulae led to the discovery of radio waves (Heinrich Rudolph Hertz, in 1888), X-rays (Wilhelm Conrad Roentgen, 1895), and radioactivity (Antoine Henri Becquerel, 1896).

All this resulted in a new definition of the kinds of radiation given off by objects in nature, including objects in space. Soon instruments would be built that could measure this radiation. These became the principal tools for studying the planet, until the space age in the mid-1960's, when another revolution took place.

3 Jupiter with the Naked Eye

It was natural for Galileo to aim his first telescope at Jupiter early in his sky watching. It is the third brightest "star" in the sky; only Venus and sometimes Mars outshine that giant world.

Galileo was only the first. Because of Jupiter's size, and the fact that its bright side shines toward us for months at a time, it became one of the astronomer's favorite objects of study.

It soon became apparent that Jupiter was a world unlike Earth and unlike Earth's closest neighbors. Astronomers theorized that Mars and Mercury, like the moon, must be rocky and crater strewn. They could see a thick cloud bank covering Venus and guessed that it too was a world of rock.

But Jupiter was different. If there was a solid world beneath the clouds, there was no evidence of it. Jupiter seemed to be a "gas giant," a world made up mostly of gas or liquid. Saturn, Neptune, and Uranus seemed similar. No one could see enough of Pluto to make any judgment.

So astronomers began to divide the worlds of the sun into two classifications: the rocky ones, like Earth, were called terrestrial; the liquid or gaseous ones, like Jupiter, were called Jovian.

Astronomers could determine many other things about

Jupiter with their earthbound instruments. When the space age finally permitted actual exploration, the explorers were not being sent to a completely unknown world. There were many questions to be answered and some mistakes made, but by and large Jupiter's measure had been taken.

Jupiter is not round. It bulges in the middle. Astronomers have been able to measure its diameter at the equator as about 143,000 km. The diameter from pole to pole is about 124,000 km, a difference of 19,000 km. Its surface area is a whopping 62 billion square km, which is 1,312 times that of Earth. But its mass, or weight, is only 318 times that of Earth. The planet is made up mostly of the very lightest elements, hydrogen and helium, and is not very dense—about as dense as earth water.

Jupiter behaves in the sky in a fairly common way. Like all of the planets, it follows a somewhat elliptical course around the sun, but as with all the planets except Pluto, the course is not really very elliptical. At its greatest distance from the sun, Jupiter is 815,620,000 km. At its closest it is 739,980,000 km. Its average distance from the sun is 779,410,000 km, and the variation from the round is only about 10 percent.

A Jovian year takes about 11 years, 11 months. But a day is only nine hours and fifty minutes or fifty-five, depending on where you measure. That whirling speed accounts for the bulge in the planet's middle. Astronomers have clocked the speed of the top of Jupiter's cloud cover at 35,400 km/hr at the equator, which is 22 times faster than the Earth spins.

Besides helium and hydrogen, scientists have identified deuterium (the heavy form of hydrogen), methane, and ammonia in the clouds.

The cloud layer visible to the telescope has always fascinated scientists. Jupiter is banded with bright salmon and red, gray and blue stripes with darker-gray areas near

Jupiter with its red eye

the poles. The larger reddish stripes are called zones, and the larger bluish stripes are called belts. Each has a name. The southern hemisphere has a different pattern from that of the northern.

Everyone believes that those are layers of clouds, and that the outermost layers, the ones visible to us, are quite cold, −145° C, although temperatures of a terrestrial spring have been recorded in some parts. The tops of the clouds are probably ammonia and crystals, with some methane in the polar regions.

One of the best-known features in the solar system is the Great Red Spot, found in the Jovian southern hemisphere. It was first seen as early as 1631, and it fades and darkens in 30-year cycles. In 1878 it was described as being very conspicuous, a dark red. It was measured at about 48,000 kilometers long and 11,000 kilometers wide, greater than the surface of Earth. It has faded and darkened several times since then, becoming a pale orange or sometimes completely disappearing or deepening into red. It drifts slowly through the cloud tops.

Exactly what causes the Spot is unknown, but two theories have emerged. One is called the "raft" theory, which says the Spot is something solid or partially solid floating on top of the clouds. The change in color is explained by its sinking or rising in those clouds.

The other theory says it is a "Taylor Column," a column of stagnant gas caused by some object on a hard surface. The wind deflects around that object, say a very tall mountain, eddying around a column of gas and thus creating the Red Spot.

The Red Spot is not the only startling feature found in Jupiter's cloud cover. Between 1901 and 1940 there was a giant storm found in the southern hemisphere at about the same latitude as the Spot. It eventually faded away and has never been seen since.

Several identifiable weather currents appear in the clouds. One, the Great Equatorial Current, is gigantic, sweeping around the Equatorial Zone at speeds of 410 km/hr, faster than the clouds on either side. There is another current that just sweeps around the Red Spot.

This constantly moving, turbulent weather is very similar in ways to the Earth's atmosphere, and scientists believe that understanding the Jovian weather pattern will help uncover some mysteries about ours.

If there is gravity on the surface of the planet, it would be about 2.5 times stronger than Earth's. It would actually vary depending on where you are standing. A 68-kilogram man would weigh 161 kg at the equator, and 186 kg at the poles, a difference of 25 kg.

Like Earth, but unlike every other planet, Jupiter has a magnetic field. Scientists know this because the planet is a huge radio transmitter, sending out very high frequency (decimetric) radio waves. The only explanation for those waves can be that a field of trapped electrons is roaring around the planet. When electrons are going fast enough to approach the speed of light, they transmit decimetric waves. Actually the process is more complicated than that, but the observation of the magnetic field, made by radiotelescopes, is quite accurate.

Those same radiotelescopes indicate that the magnetic field is doughnut shaped and dipolar, which means it has a north and a south pole, just like Earth, but 20 times stronger and perhaps a million times denser.

Those decimetric radio waves are not the only transmissions coming from Jupiter, the busiest radio station in the solar system. The planet also sends out radio waves produced by the motion of the molecules in the atmosphere caused by temperature variations, generally waves 3 centimeters long. It also emits huge waves called decametric, tens of meters long, which are probably caused by mind-boggling bolts of lightning in the atmosphere.

The decametric radiation seemed to be associated with one of Jupiter's known moons, Io, the closest to the planet. Scientists speculated that Io was triggering the lightning bolts and was acting like an on-off switch to the Jovian radio broadcasts.

Jupiter, in fact, radiates about 2.5 times the amount of energy it gets from the sun. One of the questions scientists wanted answered was where that energy was coming from. It is so strong that the temperatures of the clouds in the shadows produced by the moons is actually warmer than the sunlit areas adjacent to them. And Jupiter only gets 4 percent as much of the sun's energy as the Earth gets.

Until very recently it was believed that Jupiter had 12 moons. Some of them were, fittingly, giants, the size of small planets. The largest is Callisto, which is about 5,180 km in diameter. Close to it in size is Ganymede, which is 5,150 km in diameter. The planet Mercury is only about 4,800 km in diameter, so Jupiter has moons bigger than some worlds. They are about half the size of Earth, which has a diameter of 12,753.594 km.

The other giants are Io, 3,715 km, and Europa, 3,135 km. Those four moons are now called Galilean because they were the ones discovered by Galileo.

The smaller of the dozen known moons were Ananke, 22 km, and Lysithea, 24 km.

Amalthea is the closest to Jupiter, skimming 106,000 km above the cloud tops once every 1½ days. The farthest out of the moons, Sinope, is 23.7 million km from Jupiter.

One odd thing about the four outermost moons, Ananke, Carme, Pasiphae, and Sinope, is that they circle Jupiter in the wrong direction, going in exactly the opposite direction for all the other moons. Astronomers think that those moons may have been wandering asteroids, captured in Jupiter's gravitation force and doomed to live forever the uneventful life of a moon.

Their distance from Jupiter actually puts them somewhat within the influence of the sun's gravity, and the result is an orbit consisting of very long loops. In fact one moon, Pasiphae, was "lost" soon after its discovery and was reported "missing" for about 20 years until it was "found" again in 1940.

It takes more than 700 days for the outer moons to make their circle of Jupiter with Sinope sailing along for 758 days before completing its "annual" journey.

Some of the moons are odd in other ways. The oddest is perhaps Io. For 10 minutes after Io appears from behind

Jupiter's shadow, it glows fiercely. Scientists think this is evidence that Io has an atmosphere. When Io is in the shadow of Jupiter the atmosphere grows so cold that snow falls onto the surface. But as soon as the moon hits the sunlight, the snow melts and reevaporates into the atmosphere. Moreover, the color when Io shines is bright orange, for reasons not yet understood, and the shine makes Io the brightest object around in terms of reflected light. It reflects more sunlight than any other object in the solar system.

Two other of Jupiter's miniworlds, Europa and Ganymede, are believed to have water ice on their surfaces.

One of the great mysteries about Jupiter is the greatest mystery of all—the possibility of life. It would not be walking-around-singing kind of life, but primitive organisms floating around in the turbulent atmosphere.

The reason that the possibility of life is even being considered by scientists is the known makeup of the Jovian atmosphere. There are water droplets in there. There are probably oxygen molecules also. Some portions of Jupiter's atmosphere have temperatures about as high as the room you are sitting in, so some of that water is in liquid form, and some is ice. Other components of the atmosphere include ammonia, methane, and hydrogen.

Scientists believe that when the Earth was very young, those same components existed in the air. With the help of energy in the form of ultraviolet radiation and huge bolts of lightning, that primordial "soup" created the chemical ancestors of biologic life—amino acids. If life could have evolved on Earth, it would be irrational to think it could not have happened on Jupiter as well. All the ingredients of the recipe are there.

Scientists have suggested that the Jovian atmosphere is a giant farm, turning out vast amounts of chemicals that can support life, life without photosynthesis, or sun power. Any life that evolves would be floating free in the atmosphere and would probably be like bacteria or simple plants. One science fiction writer even imagined huge, balloonlike creatures soaring around on air currents, with tentacles dangling down to catch food, something like a Jovian jellyfish. No one can call that impossible.

The atmosphere is the most intriguing region of Jovian study. It was generally believed that the cloud tops were ice-cold ammonia crystals. Below those crystals, somewhat warmer, were droplets of ammonia and below that, ammonia vapor. The deeper down you explore, the warmer things get, scientists believed.

Below the ammonia vapor, water would go through the same cycle—ice crystals, then water droplets, then water vapor. Under the vapor would be Jupiter's solid surface (if it really has one), or oceans of liquid hydrogen (if it does not). Under those oceans is something scientists can only imagine but cannot really explain very well—metallic hydrogen. The element we know as a gas or a supercold liquid would form a solid under the high gravity that Jupiter generates.

If Jupiter has a solid core at all, it is probably made up of some rocks and mineral elements, squeezed to a mass ten times that of Earth.

The magnetic field of Jupiter has always puzzled scientists. Earth's field is produced by what is called the Dynamo effect. The core of the Earth is something like hot slush, and it sloshes around, producing magnetic waves, something like a generator producing electricity. Jupiter's field could be produced the same way with a fluid core splashing around at temperatures of about 10,000° C. Another theory is that the magnetism is the result of the interaction of the various layers of the atmosphere. No one knows for sure.

That, then, is the picture of Jupiter as it was drawn from long-distance observation. It began with the naked eye, progressed to telescopes, and eventually instruments that measured every known form of radiation.

The challenge that remained was an actual visit. Since it will be generations before humans can actually go to Jupiter, and even then they probably would not be able to get very close, the only alternative was a robot.

Designing, building, and sending off such an automatic explorer was a mind-boggling task.

4
The Pioneers

We don't know about the first humans who looked up at the sky and wondered about the heavens. We cannot identify the first human who yearned to fly into the night sky and see the stars close-up, or greet the moon, or chase the sun.

It must have been frustrating to be an astronomer, to use a gigantic telescope and look at the mountains and plains of the moon and think: I can look at them so closely they seem quite near. But I can never go there. I can only look.

Some dreamers, of course, were not satisfied to just look. They envisioned humans striding the lunar craters, hiking along the Martian canals, or trekking through steaming Venusian jungles. They were immediately written off as nuts.

Some of the "nuts" were more serious. Men like the Russian Kostantin E. Tsiolkovsky, the American Robert Goddard, Germans like Hermann Oberth and, later, Wernher von Braun, did not think it unrealistic to imagine rockets capable of carrying men into space. Despite the frequent ridicule or neglect of the Establishment, these men designed, built, and fired small rockets which developed technology needed for space travel.

On October 4, 1957, the Soviets brought the dream of

Oberth and Goddard a giant step closer, when they launched the first space satellite, tiny Sputnik I. It circled the Earth, beeping its presence. It provided no information of any use, but was the first practical demonstration that man-made objects can be sent into space. The United States was caught by surprise (actually, the Soviets had announced that they were going to launch a satellite months previously, but no one in Washington with any authority believed them) and Von Braun, now working for the U.S. Army, was sent into the breach to catch up.

The Soviets had captured a number of German rocket scientists, but most of them, led by Von Braun, made sure they were recaptured by the Americans. It soon became a race to see whether the Germans in the United States were better than the Germans in the U.S.S.R. Eventually, the scientists in America won.

By the early 1960's it had been proven that humans played a role in space. Yuri Gagarin went into orbit on April 12, 1961, Gherman Titov traveled 700,000 km into space four months later, and on February 20, 1962, John Glenn went 130,000 km in an American rocket.

President John F. Kennedy had announced that Americans were going to land on the moon by the end of the decade. Supporters of space flight raved about the advantages of space technology to the man in the street—weather and communications satellites, and space exploration.

In 1960 the first weather satellite, Tiros I, had actually been launched and sent back dramatic pictures of Earth's cloud cover, giving meteorologists a star's-eye view of weather patterns. The first weather satellite, however, was a milestone for another reason.

Men like Goddard had dreamed of sending humans along in rockets, in part because they could not think of any other way to get back information. The art of telemetry, in which information is gathered by machines and transmitted over a long distance, had not been developed. Goddard, in fact, thought at first that the earliest lunar explorers would announce their safe landing by shooting off a flare.

But by the end of World War II (Goddard died in 1945), sophisticated radio and recording equipment made the idea of lunar flares obsolete. In 1951 three Englishmen, K. W. Gatland, A. M. Kunesch, and A. E. Dixon published a paper in the *Journal of the British Interplanetary Society* first proposing man-made satellites. In 1953, S. Fred Singer of the University of Maryland proposed MOUSE (Minimum Orbital Unmanned Satellite of the Earth), which would use solar batteries to power a radio transmitter.

Tiros did exactly that, living proof that the concept would work.

It was logical to believe that if you could orbit a "talking" satellite, you could send off a "talking" space probe, a robot that would go where humans could not reasonably expect to go in the foreseeable future to do the observing for earthbound scientists.

The Soviets began such exploration before the 1950's ended, sending up Luna I ("Metchta" in Russian) on January 2, 1959. It flew to within 7,500 km of the lunar surface, sending back information on space in that region. Scientists had, for the first time, a mobile laboratory chirping back data from near another heavenly body.

In the early spring the United States sent up Pioneer IV, which got to within 60,015 km of the moon and was tracked a few thousand kilometers farther than Luna I.

The Russians upstaged Pioneer IV with their Luna II, launched on September 12, 1959. It weighed 1,360 kg and carried 389 kg of scientific instrumentation. It crashed (deliberately) into the moon on September 14. Luna II sent back radio signals up to the moment of impact and recorded the fact that there was no magnetic field around the moon. It was probably the first piece of solid scientific evidence about a heavenly body ever produced by a spacecraft.

The Soviets outdid themselves a month later by launching Luna III on October 4. It was about the same weight as Luna II, but had only 278 kg of scientific equipment, including cameras. It circled *behind* the moon, getting to within 7,035 km of the surface. Luna III sent back Earth's first look at the far side of the moon, the side that is always facing away from Earth, and thus has never been seen before. It showed a surface not nearly so crater scarred as the near side.

More important, Luna III proved, if proof was still necessary, that small, unmanned spacecraft could fulfill astronomers' daydreams. Keep in mind that Luna was launched before the electronic revolution and before miniature circuits—in fact the Lunas helped push this technologic advance. The ability of these early robot explorers was quite limited. Solid-state electronics and the transistor would turn them into scientific wizards.

Eventually both countries made "soft" landings on the moon, and both soon turned to other planets for exploration.

The Soviets had tried and failed to send spacecraft to Mars in 1960. That same year they successfully launched a spacecraft toward Venus. Unfortunately its radio quit and nothing of value was returned, although the spacecraft came to within 100,000 km of that cloud-enshrouded world. The same thing happened in 1962 with a Mars explorer, Mars I, which was one of the first spacecraft to carry television cameras. In 1960 the United States achieved a major victory in the space race with a successful deep space mission, using a spacecraft dubbed Pioneer V. The first to study interplanetary space, it was tracked as far as 36.2 million km, out toward the orbit of Venus.

After a failure to actually fly by Venus, Mariner II pulled off another major coup. Launched on August 27,

1962, it flew to within 35,400 km of Venus, sending back data from 56.3 million km. The large spacecraft weighed 202 kg, including 18 kg of scientific equipment. Among the new things learned from that mission is that Venus is quite a terrible place, with temperatures on the surface reaching 370° C, mountains, and an unbroken cloud cover.

The record of unmanned exploration of the planets was excellent. Inevitably, space scientists thought that they ought to try for Jupiter and the outer planets.

It was determined that two spacecraft would do the job. The first would be the pathfinder, the scout. Astronomers had several good reasons to think that the trip was a dangerous one for any spacecraft, manned or otherwise. If the first one ran into trouble, the flight of the second could be altered to avoid the danger.

The spacecraft were given one of the most successful names in the annals of the National Aeronautics and Space Administration—Pioneer.

The first Pioneer spacecraft had been one of the real pioneers, launched in October of 1958. It had set a record by reaching 113,700 km into space, the deepest penetration at the time. Pioneer II was a failure because the last stage of its three-stage rocket did not fire.

Pioneer III, in 1958, discovered the outer Van Allen radiation belts around the Earth. Pioneer IV was the first spacecraft to escape Earth's gravity, the first free-flying man-made object in space.

Pioneer V, in 1960, monitored the sun. Pioneers VI and VII were launched into solar orbit, the first man-made satellites of the sun. Pioneer VI went into solar orbit between Earth and the sun, Pioneer VII went into solar orbit between Earth and Mars. Pioneers VIII and IX were also solar-orbiting satellites.

The result of the solar Pioneer flights was new, previously unattainable information on such things as sun storms and sunspots.

Pioneer/Jupiter spacecraft

The task of building Pioneers X and XI, the Jupiter probes, went to TRW Systems, of Redondo Beach, California, just south of Los Angeles. TRW is one of the most experienced of the aerospace companies, having built several of the earlier Pioneers. Control over the project went to NASA's Ames Research Center at Moffett Field near Palo Alto, California, south of San Francisco.

Engineers first needed to know what the scientists wanted to find out about Jupiter. The second task was to figure out instruments that could do what the scientists wanted. The third problem was building the instruments so that they would fit neatly into the spacecraft and weigh as little as possible. Every gram of weight in a spacecraft multiplies itself hundreds of times over in valuable rocket fuel. The game in space flight is to save as much weight as possible.

Since the new spacecraft would have to go farther than any spacecraft in history, they would spend more time in interplanetary space, more time out beyond Mars where no other spacecraft had ventured. An ideal opportunity to study deep space was an added bonus to the Pioneer missions.

The scientific community, seizing on an opportunity they once could only dream about, came up with seventeen fields of study they thought Pioneer might explore for them. The spacecraft were to:

—Map the magnetic fields in deep space.
—Study the solar wind, the almost imperceptible wisps of particles streaming from the sun.
—Study cosmic rays, which are tiny electrically charged particles coming from the sun and from intergalactic space.
—See if there is any relationship between the behavior of the solar wind, the magnetic fields, and cosmic radiation.
—Find the outer edge of the heliosphere, the invisible atmosphere of the sun.
—See what quantities of free-flying hydrogen atoms exist in interplanetary space.
—See how much dust there is in space.
—Map Jupiter's magnetic field.
—Measure the kinds of atomic particles that exist in Jupiter's radiation belts and look for the Jovian equivalent of the northern lights.
—Try to find the source of both the decimetric and decametric radio transmissions from Jupiter.
—Measure the *magnetosphere* (the magnetic field) and the *bow shock wave,* which is the splash of radiation caused when the magnetosphere knifes through the solar wind.
—Take the temperature of Jupiter's outer atmosphere.
—Measure the ratio of hydrogen to helium in the atmosphere.

—Study the nature of the atmosphere.
—Measure how much light is reflected from Jupiter.
—Take pictures of the planet.
—Increase the accuracy of what we know about Jupiter's orbit and about the orbits of its moons.

All of those tasks were the kinds of things that could not be accomplished from Earth. In fact, what NASA was really asking the scientific community was: If you were going on a spaceship to Jupiter, what kinds of things would you most want to do on the trip? Scientists love questions like that.

Many, many suggestions were thrown out or simply rejected either because they were impractical or because the information gained was deemed of lesser importance and not worth the additional weight in the spacecraft. In space flight, weight is everything.

A number of very worthy suggestions were also rejected simply because of the limits to the kind of spacecraft NASA could build.

But the scientific community was in almost complete agreement that the seventeen tasks set for Pioneer would provide more information about Jupiter than had been gathered in all the previous centuries. It would be a gigantic leap in understanding that most mystifying giant world.

Thirteen experiments were designed for Pioneer X, involving many different pieces of equipment aboard the spacecraft. Another experiment was added to Pioneer XI.

Let's see how those instruments work and what they can do.

Magnetometer

Built by NASA's remarkable Jet Propulsion Laboratory at the California Institute of Technology, the magnetometer, as its name implies, measures magnetism—the magnetic fields in space. It has a sensor that rides on a boom 6.5 meters long extending from the spacecraft. That keeps it out far enough so that the magnetic field being generated by the electronics on Pioneer itself does not interfere with its readings. It can measure magnetic fields in eight ranges, or intensities, including minute ones 0.00002 percent of Earth's field. It operates either on command from Earth or on commands stored in Pioneer's computer. Specifically, the magnetometer measures the interplanetary magnetic field, the interaction of the solar wind with the planet, and Jupiter's magnetosphere, or magnetic field. Some of the data will be useful in determining the consistency of Jupiter's interior, particularly if it turns out to be liquid. The whole instrument weighs only 2.6 kg.

Plasma Analyzer

Built by NASA's Ames Research Center, this instrument is designed to study the solar wind. The instrument consists of a slit in the dish-shaped antenna of the spacecraft that is open toward the sun. Particles of the solar wind, electrons, and charged particles, called ions, enter the slit and strike metal plates that count them. The instrument records the direction in which the particles were traveling, their energy or speed, and the number of particles. It weighs 5.5 kg.

Charged Particle Composition Instrument

This instrument from the University of Chicago consists of four measuring devices that can identify all eight chemical elements running from hydrogen to oxygen, and their various forms. Two of the systems would work at Jupiter, two others (telescopes of a peculiar nature) work in interplanetary space. The whole thing weighs 3 kg.

Cosmic Ray Telescope

A slightly different device from NASA's Goddard Space Flight Center, Maryland (and how Robert Goddard would have loved that name), which monitors the cosmic rays. It can tell which of the ten commonest elements make up the ray particles and see how those rays are deflected by the sun's magnetic field. It will also take the measure of particles in Jupiter's radiation belts. This device weighs 3.2 kg.

Geiger Tube Telescope

The Geiger tube telescope will measure the characteristics of Jupiter's intense radiation belts. The instrument was designed by James Van Allen (of Van Allen belt fame) and his colleagues at the University of Iowa. The tubes contain a gas that produces an electrical signal when it is hit by ions believed roaring around in those belts. Weight is 1.6 kg.

Jovian Trapped Radiation Detector

This machine, from the University of California at San Diego, probes even deeper into the radiation belts. It produces data on the particles in the belts, including whether they spin or not, how much energy they have, and how they travel. Consisting of five detectors, it weighs 1.7 kg.

Asteroid-Meteoroid Detector

One of the suspected danger zones in the Pioneer flight is the asteroid belt, a belt of rubble and small planets that circles the sun between the orbits of Mars and Jupiter. There was fear that Pioneer might run into some of these pieces of rock and die a victim of the first mid-space collision. This detector, built by General Electric, consists of four electronic telescopes that scout around as Pioneer whizzes through space. It finds the dust, stones, rocks, boulders, and miniworlds by spotting the sunlight reflected from them. It can measure things that would be invisible to the human eye. The instrument has one weakness, it is easily damaged by high radiation and could be knocked out when Pioneer flies through the Jovian radiation belts. It weighs 3.3 kg.

Meteoroid Counter

This device measures the number of times Pioneer is hit by particles during its flight. Built by NASA's Langley Research Center, Virginia, it is a honeycomb of little gas pockets. Every time a particle hits the honeycomb, which is mounted on the back of a rear-facing antenna, it pops a chamber. This clever little gadget weighs only 1.7 kg.

Celestial Mechanics

Celestial mechanics involves measuring the gravity field of Jupiter. The system designed by the Jet Propulsion Laboratory uses the spacecraft as its instrument for measuring how much the Jovian gravity affects the predictable path of the spaceship. Radio detectors will spot the spacecraft along its flight to within one millimeter per second once every minute. A radio detector will know where the spacecraft is supposed to be, so celestial mechanics can measure even the slightest wobble caused by Jupiter or its moons. This will also enable scientists to tell the mass (or weight) of these objects to within one percent in the case of the moons. All the instruments for this experiment are on Earth, so the celestial mechanics experiment has no *weight cost.*

Ultraviolet Photometer

This instrument, from the University of Southern California, is calibrated to identify two elements, hydrogen and helium, in the Jovian atmosphere. The photometer measures how much ultraviolet light from the sun bounces off particles. If there is hydrogen in the atmosphere (and scientists are sure there is), it would reflect some of the light at a certain wavelength. If there is helium (and scientists suppose there is), it would use still another wavelength. It will also measure the helium in the intergalactic cloud our solar system is sailing through. If the big bang theory really is true, there should be a reading of 7 percent helium in the interstellar gas. Weight is 0.7 kg.

Imaging Photopolarimeter

This is a camera with a fancy name. The device, built by the University of Arizona, will provide the flights of Pioneer X and XI with some of their most dramatic moments when the spacecraft send back breathtaking close-up color pictures of Jupiter. We will discuss this instrument in detail later; suffice it to say that the whole thing cost only 4.3 kg. of weight and was worth every gram.

Infrared Radiometer

This instrument is used to find some possible solutions to the great heat mystery: Why does Jupiter radiate much more heat than it receives from the sun? Using the infrared band, the instrument finds out just where the heat is coming from, the temperature distribution, hot and cold spots, and the relative abundance of hydrogen and helium in the atmosphere. It weighs 2 kg. The experiment came from the California Institute of Technology.

Pioneer XI, the sister ship of Pioneer X, sits ready to go

Occulation Experiment

This is another no-weight experiment from the Jet Propulsion Laboratory. The purpose of the experiment is to watch Pioneer as it disappears behind Jupiter. The time and nature of the disappearance will tell scientists where the top of Jupiter's atmosphere is. It will also measure how far radiation is bent by the atmosphere. Experiments like this can be done by observing stars that pass behind the planet, but using Pioneer and its more precise radio tracking will give more accurate readings. The experiment will

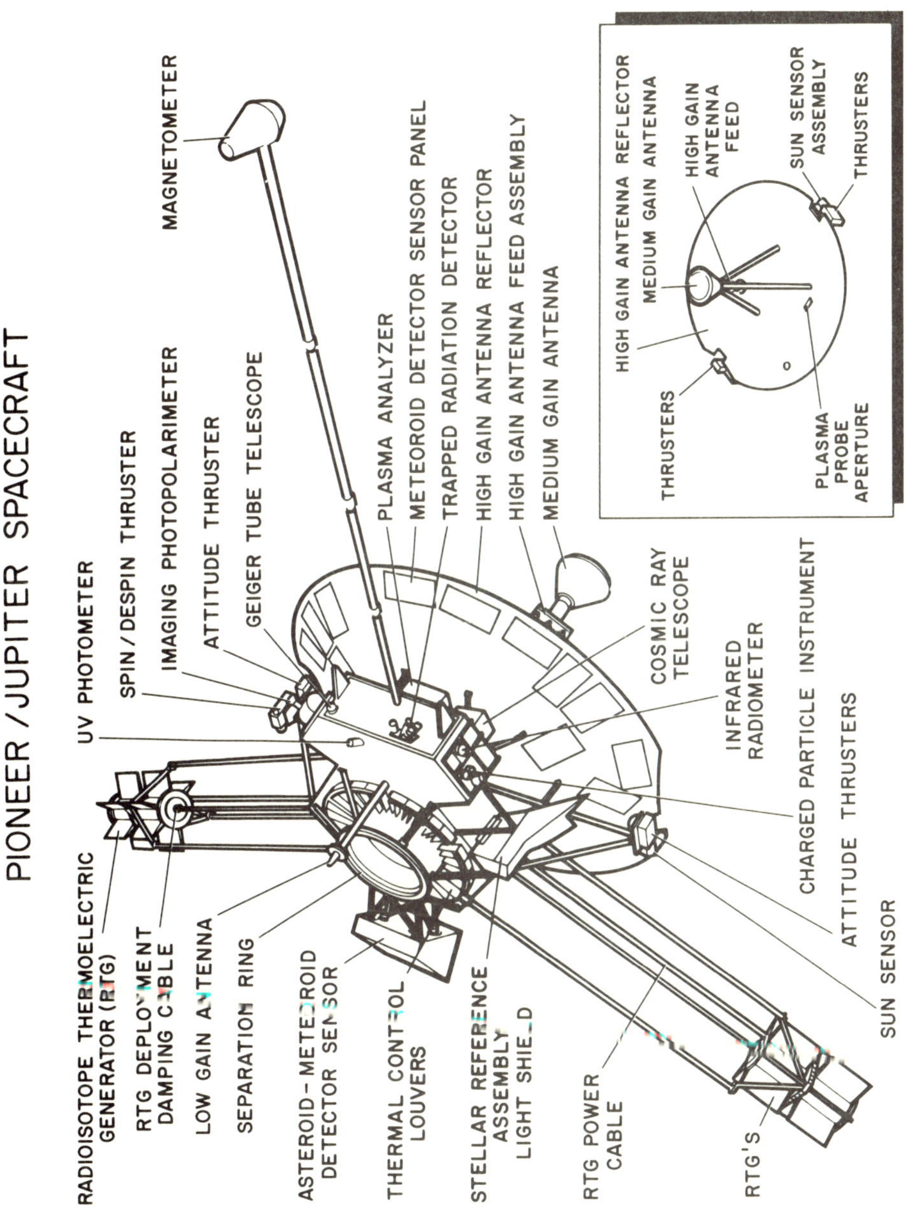

Diagram of the Pioneer/Jupiter explorers

also be tried when Pioneer passes behind Io, the moon with the suspected atmosphere.

Those are the experiments. They are notable for two things. One, they weigh very little. All the wonders of modern electronic and solid-state miniature circuitry have been used to keep their weight to a bare minimum. The entire scientific package weighs a little less than 30 kg, much less than the weight of a human being.

Two, scientists from all over America eagerly joined in designing the spacecraft.

All the equipment was crammed into a fairly small package. It is 2.9 meters long and 2.7 meters wide. The whole craft weighed 270 kg at launch.

The forward tip of it is the dish antenna, aimed so that it will usually face Earth. The equipment compartment, behind the antenna, is in the shape of a six-sided box, 35.5 centimeters deep. Each side is 71 centimeters long. Most of the scientific equipment is in a smaller box, 35.5 centimeters deep alongside the main compartment. The other instruments ride outside the spacecraft, some of them on long booms. The spiderlike craft is also cluttered with three antennae sticking out from the main body.

The whole thing is made basically of aluminum, either as panels and beams or as a honeycomb.

To maintain the stability of the spacecraft, scientists designed it to spin slowly on its axis 4.8 times every minute, moving with an antenna always facing toward Earth.

Earth-based scientists always know where Pioneer is by listening to the Doppler shift. This is a phenomenon that

The "ears" of Pioneer. Huge 64-meter antenna at Goldstone, California—one of three that listened to the tiny 12-watt radio of Pioneer from as far as two billion miles away

causes radiation waves to stretch when they come from an object that is moving away from an observer. When an object moves toward the observer, the radiation waves shorten. If you have ever listened to the whistle of a train passing, you notice that the sound rises in pitch until the moment the train passes, then it lowers. That is the *Doppler effect.* Ames scientists can calculate where the Pioneers are heading with a formula using the Doppler effect of their radio transmissions, which are channeled along a tight beam from the spacecraft's antenna toward Earth.

Pioneer is maneuverable, containing its own propulsion system. It has its own attitude control system as well, aligning itself during its flight at the correct angle to keep one of its three antennae facing in the proper direction, or to point instruments the right way, or to maintain the spin.

Six thruster, or rocket, jets point in various directions to accomplish changes in velocity and attitude. They can be ordered from Earth to fire or can be fired in a sequence contained in Pioneer's computers.

Because the Pioneers will be flying far from the sun and will receive only 4 percent of the sun's energy in the vicinity of Jupiter, solar energy panels like the kind used on other spacecraft won't work. The Pioneers are powered instead by four nuclear power plants, called SNAP-19's,

[DIAGRAM—PAGE 49]

The message to aliens, the Pioneer plaque. The plaque contains drawings of a man and a woman. The man's hand is raised in the sign of peace (which also shows how our arms and hands work). Both figures are shown against a sketch of Pioneer to indicate scale. In the upper left is the chemical sign of the hydrogen atom, the commonest in the universe. Below that is a cosmic roadmap showing our sun relative to 15 pulsars, or stars that make rhythmic radio transmissions. This would permit aliens to pinpoint our solar system and, since pulsars slow down eventually, tell when the spacecraft was launched. In the lower left is a drawing of our solar system showing which planet Pioneer came from

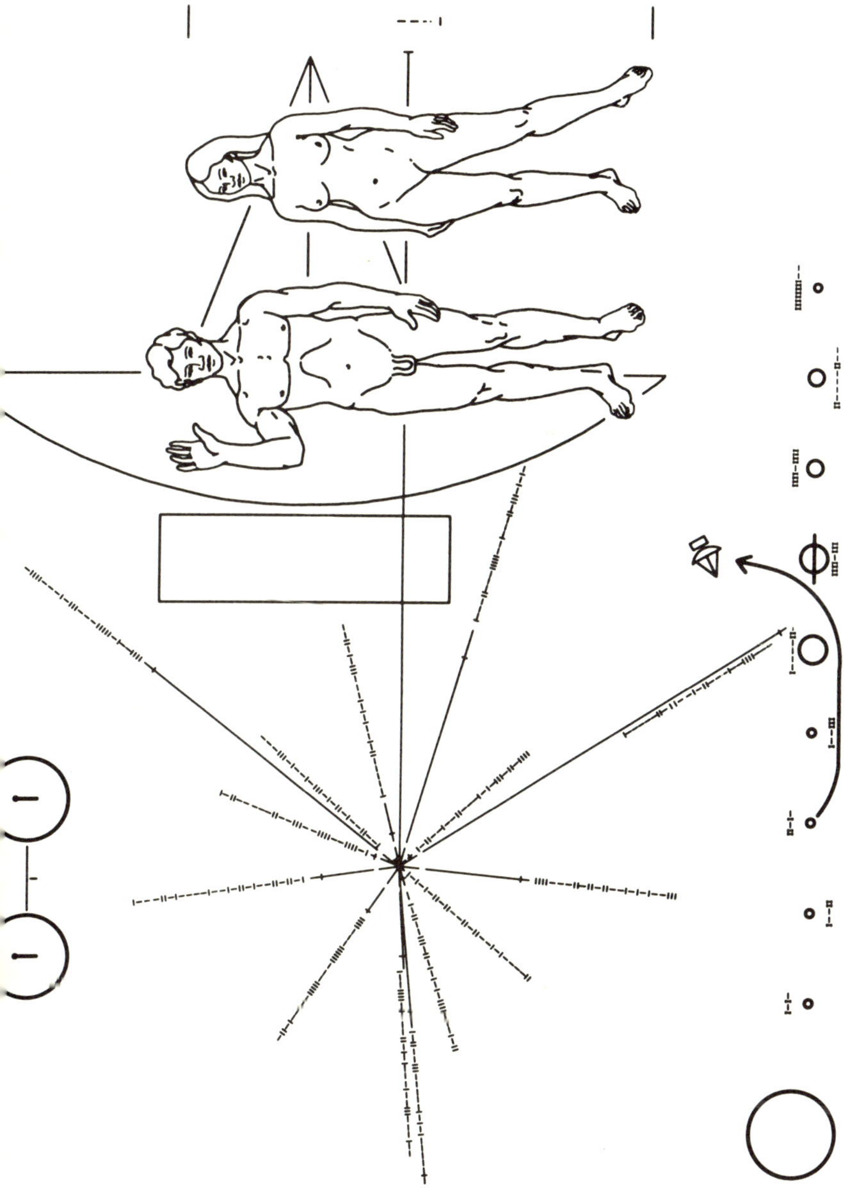

that convert nuclear energy into electricity. The SNAP's, also known as radioisotope thermoelectric generators, or RTG's, are positioned on trusses away from the spacecraft so that any radiation produced by them does not interfere with the delicate scientific instruments.

Pioneer has two radio receivers, two transmitters, three amplifiers, and three antennas. Like almost everything on board the spacecraft except the scientific equipment, the system is redundant, which means that if something fails, a backup system will jump in to do the job.

Pioneer has its own computer—also with redundant parts. Earth can send a total of 222 commands out to Pioneer. The messages, coded in *binary digits* (0, 1)—the language of digital computers—go up at the rate of one *bit* per second. A bit is a single electronic impulse. The computer gets either a 0 or a 1 by radio every second. Every message consists of 22 bits, so it takes 22 seconds to send up a command. Some of the commands control the experiments. The rest run the spaceship.

The computer can store five commands for future use if the ground-based technicians wish. The computer also has a system that verifies the accuracy of every command, thus eliminating errors.

Also the computer can store information up to 49,152 bits' worth, from the experiments and from its own operations. It can store and transmit simultaneously and can remember information faster than it can be sent. Transmission of the data can run from 16 to 2,048 bits per second by radio to Earth. When one of the more demanding scientific instruments, such as the camera, is storing information, other instruments are turned down to allow as much capacity in the memory as possible.

Temperatures inside the spacecraft are controlled at between −25° and 40° C. Anything beyond that range would damage the delicate electronics on board.

All in all, Pioneer is a marvelous device, designed to go

where no humans have gone before—the eyes and ears of science exploring strange worlds.

The public relations men from TRW noticed that the Pioneers would keep going after the Jupiter flyby, eventually leaving the solar system and entering interplanetary space. They would be the first man-made objects to travel out there. What if some intelligent creatures from another star system found one of them? Would he, she, or it know where Pioneer came from? Would the alien know what kind of creatures built it? Not likely.

So, with NASA's permission, Dr. Carl Sagan of Cornell, the inventor of the science of exobiology (the possible biology of extraterrestrial creatures) was asked to design a plaque giving greetings to any alien finding a Pioneer in the centuries hence. With the help of his artist wife, Linda, Sagan designed such a plaque, and it was built and put on board Pioneer X just three weeks before launch. It contained a drawing of two naked humans, a man and a woman, and a number of scientific notations that Sagan believed any intelligent space-flying creature could understand and perhaps trace back to our solar system and our world.

If any creatures find it, they will be very lucky. Pioneer X and XI will not enter any star system for another 10 billion years. It will take 80,000 years to travel the 4.3 light-years to our nearest stellar neighbor. Perhaps a wandering spaceship will spot the space probe and amble over to take a look. The creatures will probably think it a childish device built by a very primitive civilization.

They will be right.

5
Smile

Everyone was anxious to see the readings the first Pioneer would send back from its delicate instruments. But as vital as those readings were, they would be quickly overshadowed by the pictures Pioneer X sent down. Not even the persons who invented Pioneer's "camera" could have predicted what happened during the Jupiter flyby, because their machine worked far beyond anyone's expectations.

The cameras that Pioneer X and XI took with them to Jupiter were called imaging photopolarimeters or IPP for short. They really were not cameras in the technical sense, and they needed some help from computers and from the artistic senses of some Arizona scientists. There was, to be honest, a little hokum involved.

The IPP was invented at the University of Arizona in the mid-1960's. It was supposed to go on an earlier spaceship, but NASA rejected the design. A graduate student at the university, who was working on his master's thesis, modified that design somewhat and it was submitted for the Jupiter trip. This time NASA accepted it. The IPP is a complicated system. NASA and the designers are quick to point out that the result is not really photography and is not really television either. They call it "imaging," which, as a matter of fact, isn't really a word, but space scientists are forever inventing words to describe what they do.

The Pioneer spacecraft spins almost five times a minute during its flight. As it does, a telescope attached to the IPP is aimed in the direction of the planet. Every 1/1000th of a second, the telescope measures the intensity of the light it sees in two colors, red and blue. Each measurement is called a *pixel*, and each pixel is recorded by the IPP in the order the telescope makes it.

The intensity of the light it records is converted into a number, a number between 0 and 63. The darkest possible light it records is given a 0. That is the darkest black. The brightest light it sees is given a 63. That is almost pure white. After each roll of the spacecraft the IPP will have recorded 508 pixels, which are stored in the Pioneer's computer memory. That means that with each roll of the spacecraft the IPP has recorded a line of 508 numbers, ranging from 0 to 63. Since the computer is only allowed to remember 508 numbers, it then performs what computer engineers gleefully call a "data dump."

The information is "dumped" from the spacecraft via radio back to Earth. This transmission requires a grand total of three watts of electricity from Pioneer's nuclear generator.

The stream of data from the IPP, which has been converted from regular numbers to computer language (0, 1) is received by NASA's Deep Space Network of radio receivers located at Goldstone, California; Madrid, Spain; and Canberra, Australia. From those antennas the information is relayed to the Jet Propulsion Laboratory in Pasadena, and finally on to Ames Research Center at Moffett Field.

Each 508 pixels, transmitted as one bunch of numbers, represents one horizontal line in a picture. As the Earth stations receive this line, it appears on a television screen as a line of varying shades of light that takes twelve minutes to crawl from one side of the screen to the other.

In effect, however, what NASA is getting is two pictures, one blue and the other red. These pictures are not very pretty and not very valuable either, so back in Arizona copies of each are placed into a computer for a little wizardry.

The scientists cannot simply combine the two pictures, the blue one and the red one. That would make Jupiter look purple. We know enough about the planet from telescopes to know that Jupiter is not purple. It is in reality a series of yellow, red, blue, green, orange, and gray bands and swirls.

So to make the pictures look realistic the Arizona scientists do two things to them. One process is called *enhancement,* the other is called *cosmetics.*

Enhancement makes the pictures easier to see by greatly increasing the contrast, something like turning up the contrast knob on your television set. In the case of Pioneer, the turnup is done with numbers. Let's suppose a certain picture from the IPP has a darkest shade of 16 and a brightest shade of 28. Instead of simply accepting those two numbers the computer expands the image. It will assume that 16 is really a 0 and the 28 is really a 63, with everything else filling in between, so instead of reading 12 different shades it reads 64, five times as many.

Scientists at Arizona are convinced that what they get out of this is a fairly accurate representation of what Jupiter would look like if there was a human aboard Pioneer looking out the window.

Cosmetics is where the hokum and artistry come in. Remember that the IPP sees only two primary colors, red and blue. Obviously, a lot of the real colors get lost that way. There are three primary colors. One of them, yellow, is just missed by the IPP. So the computer technicians at Tucson add it themselves.

They know what Jupiter looks like through telescopes. They fiddle with the computer, adding or subtracting yellow until the end result looks just like the colors seen in telescopes. This adds realism, according to Dr. William Swindell, one of the IPP developers.

Cosmetics are also used to fill in the gaps caused by lapses in the radio communications. When a gap appears, the Arizona scientists add what they think fits in the hole. That process improves the quality of the pictures. It has no real scientific value but it makes the resulting photographs made from the IPP images seem more complete.

Distortion is also eliminated by the computers. As Pioneer draws closer to Jupiter the image sent back to Earth is distorted. One reason is the curvature of the planet. Another is that the spacecraft and the planet are moving faster and faster, relative to each other. The Arizona computers, using precise mathematical formulae, will predict how much image distortion is coming through on each picture and will correct for it.

Besides the obvious advantage of taking pictures from close up, the IPP was designed to improve our knowledge of Jupiter in three ways:

—Pioneer will be able to view all of the planet, not just the face toward Earth.

—Pictures taken from Earth are distorted by Earth's atmosphere. This is totally unavoidable. Pioneer's images will not have this problem.

—Instead of taking the pictures all at one time, the IPP takes a series of connecting images to produce each picture. This makes for greater clarity because pictures that are assembled bit by bit are clearer than ones that are not (if the bits are small enough). This is like using fine-grained film.

One more thing. If the IPP sends back blue and red, and the computer adds yellow, how do you get white? If you mix red, blue, and yellow paint, the result would be a dark, muddy blob. But if you mix red, blue, and yellow *light*, you get pure white. White is not an absence of color, but rather it is a combination of all the primary colors. That is how color television sets work, and that is the way the scientists planned on getting the color white from Jupiter.

According to the flight plan, the IPP on Pioneer X would get close enough to the planet to pick out objects 28 km by 193 km, about the size of the State of Connecticut. Considering how much larger Jupiter is than Earth, that is doing very well.

The plans also included pictures of each of the four Galilean moons during the flyby.

The IPP was scheduled to be turned on a week before *periapsis,* the closest encounter with Jupiter.

By the time the IPP was finished with its work, one of its developers, Tom Gehrels, would call its performance a "miracle." Scientists all over the world would agree.

6
Deep Space

Cape Canaveral is a very peculiar place. It is built on what was once a swamp. There are alligators still living there, and subtropical birds occasionally are frightened into flight by huge space rockets. Pelicans skim the surface of still waters, while men drive by on their way to the moon. Some of the most important events in human history have begun there, yet it is one of the few places on Florida's east coast where, until recently, you could walk on a beach for nearly a mile and not see another person, not even a tourist.

It seems, on all counts, to be a strange place to begin a trip to Jupiter. But toward the end of February 1972 that is exactly what was going on there.

The Pioneer X spacecraft sat covered with a metal shroud on top of an Atlas-Centaur rocket. The Atlas-Centaur was one of the most reliable of all of NASA's launch vehicles. It is a two-stage rocket, liquid fueled. The first stage, which actually has two sections, builds up to a total of 176,000 kilograms of thrust, and the second stage builds to a thrust of 13,600 kg. It was powerful enough to lift 1,244 kg to escape velocity—the speed needed to break free of Earth's gravity. It stood 20 meters high and was 3 meters in diameter at its widest.

What most people do not know about large space rockets is that they are very thin-skinned. Most of the strength in the walls of the rocket come from the fuel inside. Empty, the walls are very fragile; full, they are still quite vulnerable.

And vulnerable was what that Atlas-Centaur was when launch time approached. Weather balloons were reporting that the wind high above Cape Canaveral (then known as Cape Kennedy) was really roaring. Ground technicians were afraid that if they launched the rocket through those winds, it would simply break in two. So they waited.

Then, when the winds died down, an electrical storm moved in. Florida has spectacular electrical storms, especially in the spring and summer. They involve whole walls of mean black and purple clouds lighted by gigantic bolts of soundless lightning. One of those walls moved across the Cape and one of those bolts of lightning hit near the launch control building, knocking out power. Another delay.

Finally Earth's heavens relented and on the night of March 2, 1972, the Atlas-Centaur, carrying mankind's first Jupiter probe, blasted off the concrete and sand. It went faster and faster, dropping off its bottom stage, hurtling farther and farther out. The top stage shut off and blasted free, and the now-unshrouded Pioneer silently whisked away from Earth and on its perilous journey.

It was a very perilous journey, and the people who sent it could never guarantee that Pioneer would get to Jupiter alive and well.

During the flight there were two danger periods that greatly frightened the mission controllers. The first was

Launch of Pioneer X, March 2, 1972

UNITED
26

when Pioneer became the first man-made object to hurtle through the asteroid belt.

The asteroid belt is one of nature's great rock gardens, lying between the orbits of Mars and Jupiter. It was not until the eighteenth century that anyone knew it existed.

Two scientists, J. D. Titius and later J. Bode, suggested that if you looked carefully at the mathematics of the known solar system, the 483 million km between those two worlds was too empty. There should be another planet in between Mars and Jupiter.

A Hungarian astronomer, Baron von Zach, formed a club of "planet hunters," mostly amateurs with telescopes, to find the planet predicted by Bode's mathematics. They worked out a logical method of scanning the sky for that world, but they were not quite fast enough. An Italian, Piazzi, beat them to the discovery on the first day of the eighteenth century. Piazzi found a small world out there. He named it Ceres, and announced that Ceres was in orbit around the sun 414 million km away from it.

Von Zach's planet hunters were not completely discouraged, however. Ceres was simply too small to satisfy Bode's law (we now know Ceres is only 687 kilometers in diameter), so they decided to keep looking. They were rewarded for their efforts. In the next six years they discovered three more little worlds, Pallas, Juno, and Vesta. When they could find no more, the club disbanded. It would not, incidentally, be the last time amateur astronomers would make important discoveries. Nevertheless it was 40 years before another little planet or asteroid, Astraea, was found. Since then, every year astronomers have found asteroids, and now there are 1,700 known and plotted, and there could be as many as 40,000 substantial hunks of rock out there.

Where did they come from? It is possible, but considered unlikely, that a large world once existed out there, but something happened to it and it exploded. Another idea is

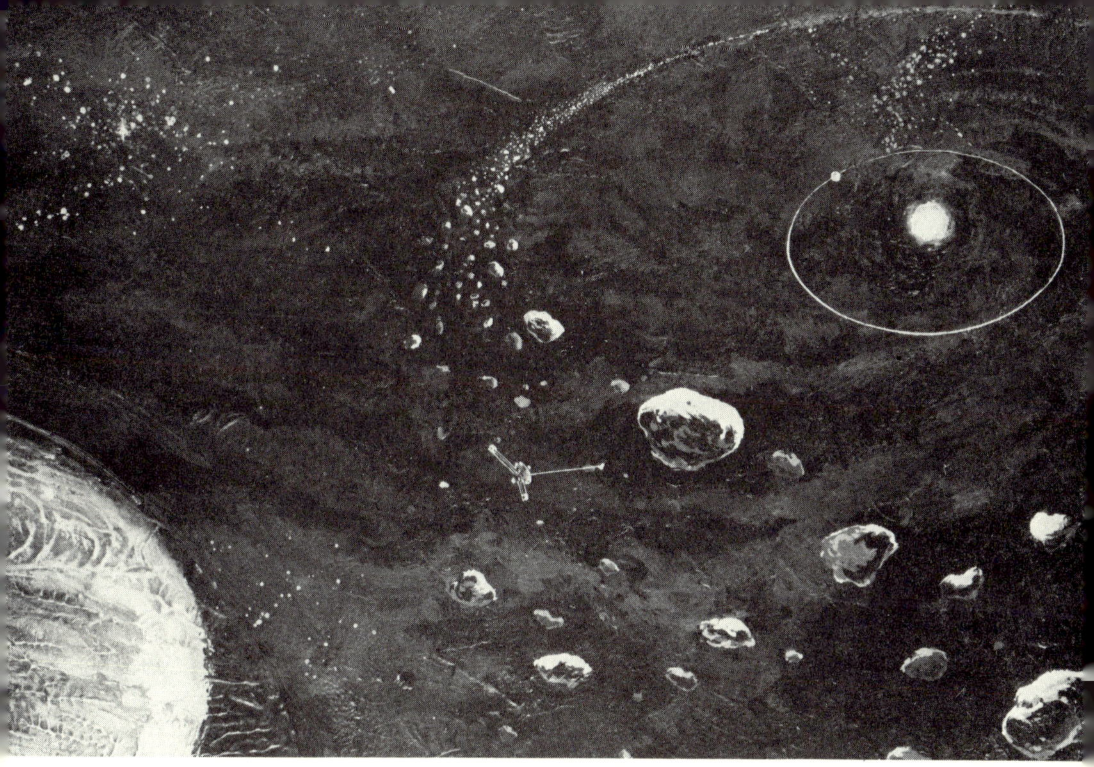

Artist's drawing of Pioneer sailing through the asteroid belt from Earth toward Jupiter

that, for some reason, the primordial cloud at the creation, which we discussed, formed large rocks, but enough of them never came together to form a real planet.

Whatever the cause, the asteroid belt is a doughnut-shaped stream of planets or asteroids, clumped in thickly populated areas separated by relatively open spaces called *Kirkwood gaps.*

Pioneer X was to fly through a Kirkwood gap. The problem was: Just how empty are the gaps? If there are hunks of rocks in the asteroid belt, why not stones or pebbles or a rain of dust? A small stone hitting Pioneer at the speeds at which the spacecraft was traveling would be disastrous, ending the mission before Pioneer ever came near Jupiter.

About six months after launch, Pioneer X whizzed into the asteroid belt. It was the first spaceship to enter the belt, and scientists were anxious to see what it would find and, of course, to see if it would survive.

Pioneer certainly survived. What it found was that those gaps in the belt were quite empty. There were no more dust particles in the belt than there were outside it. The smallest particles, those 1/1000th millimeter in diameter, actually decreased the farther out Pioneer went. The larger particles, 1/100th to 1/10 mm, were evenly distributed all the way from Earth to the outside of the belt, with no increase recorded at all as Pioneer sailed through. Pioneer found very few particles more than one millimeter in diameter, although they increased in number in the belt.

The meteoroid counter, the device with the little bubbles of gas, recorded 25 "hits" going from Earth's orbit to the inside of the asteroid belt. Inside the belt only 17 bubbles were struck. Pioneer X never went within 6.4 million km of any large asteroids.

The Pioneer equipment also recorded a phenomenon called the *gegenschein.* The gegenschein is a glow that can be seen in the night sky by the naked eye from Earth. It always appears opposite the position of the sun, and was thought to be a reflection of sunlight caused by Earth's atmosphere. Astronauts have seen it from as far as the Moon, and now Pioneer's IPP saw it from the asteroid belt. The scientists deduced from this that the gegenschein has nothing to do with Earth, but is a glow in the sky that can be seen from all over the solar system, a real phenomenon, not an optical illusion.

So Pioneer whisked through the first danger zone in six months without more than a minor scratch, if anything. It kept sending back data on interplanetary space as it zoomed toward its destiny and the place in space that really terrified its makers.

But meanwhile that data from deep space fascinated astronomers. They had never been able to get a firsthand account before.

Pioneer, for instance, told them about the severe turbulence in the solar wind, which goes as far as 510 million km out. Close to Earth the solar wind contains fast currents of high-speed particles moving at 2.52 million km/hr. Farther out into space the difference is less. The faster streams are not as fast and the slower ones are not quite as slow. The interaction between the two streams causes friction and some heat. All this was going on much farther out in space than first thought. It looked now as if the solar wind did not calm down much until well beyond Jupiter's orbit.

This far-reaching turbulence had another effect. Scientists believed that as the solar wind decreased, more high-energy cosmic rays would get through and be recorded by Pioneer. They thought the place in space where this would happen would be fairly close to Earth. They were wrong. Because the turbulence in the solar wind does not die down near Earth, there is also no marked increase in those particles which stream in from other galaxies.

Pioneer was also able to measure those particles much better than they had been measured before to see what atomic nuclei they were made of. Scientists identified sodium and aluminum for the first time and they were able to measure the relative abundance of the elements helium, carbon, oxygen, nitrogen, fluorine, neon, sodium, magnesium, aluminum, and silicon. More facts about nature marked down in the records—another tiny piece of the picture puzzle of our universe.

On November 4, 1973, Pioneer X came to within 30 days of its closest approach to Jupiter, technically known as periapsis. This meant that the period of "encounter" was about to begin.

Activity at Ames Research Center began to pick up. A large number of scientists whose experiments were primarily tuned for the encounter began to arrive. With them came the first vanguard of reporters, science writers, and broadcast technicians. There would soon be nearly a thousand of them who would file millions of words and thousands of pictures to Earthlings everywhere.

They flew into San Francisco or San Jose and gathered along the asphalt strip near El Camino Real, the Royal Highway of the Spaniards, now turned ugly by endless fast-food stores and shopping centers. Hotels in Palo Alto, Sunnyvale, and Mountain View began to fill up. Contractors, the companies that made the hardware for the Pioneer mission, began setting up "hospitality suites" for off-duty parties. Bars were filled. Everybody was going

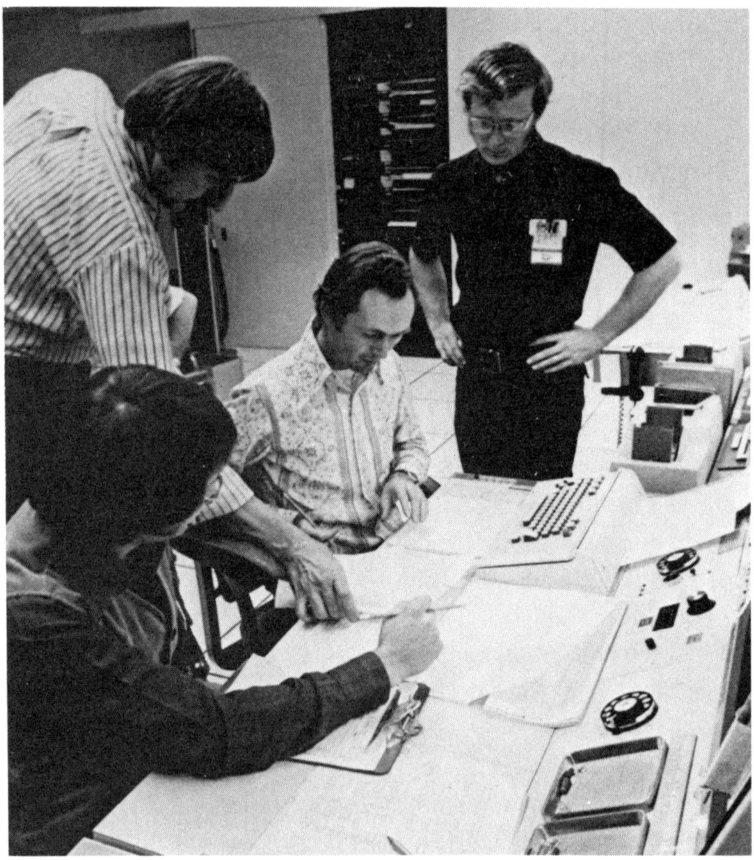

As new data comes in, the mission controllers check the computer printouts

Explaining it all to the press in the Ames newsroom. A model of Pioneer is to the right. Telephone and Telex lines link the hundreds of reporters to their newspapers and broadcast studios around the world

exploring. Once exploration meant cruising down a jungle river in the mosquito-infested tropics, or freezing to death on the Antarctic plateau. Now it meant lounging in a step-down bathtub at Rickey's Hyatt House, devouring a bag of McDonald's french fries, sitting in swivel chairs in front of computer terminals, staring at blinking lights in darkened rooms. It meant being interviewed, racing along the freeway to get to meetings, or playing practical jokes on the gullible. One television reporter almost went on his network to report the discovery of the thirteenth moon of

Jupiter—a hoax. He was stopped in time. Shortly afterward the thirteenth moon of Jupiter *was* discovered, but the once-burned reporter refused to air the story.

But out in space little Pioneer X was caught in the grip of Jupiter's mighty gravity. It was now the fastest man-made object in history, being pulled down toward the giant planet by its great gravitational field. By November 17, it was going at 33,800 km/hr and would gradually speed up to 129,000 km/hr. The IPP began snapping pictures. By November 14 it had already taken and sent back 10, showing a steadily growing world of colored bands and zones.

The danger was now approaching. Jupiter had gigantic magnetic fields. There were areas of intense radiation and electrical discharges far stronger than anything experienced near Earth. Just how strong these are could not be known until Pioneer actually encountered them. There was great fear that the radiation would "blind" the delicate electronic equipment on board, making Pioneer a useless hulk of very expensive metal. It was also possible that the radiation would alter or distort some of the readings, trick the computer into giving out false orders, or even turn it off. Some scientists seriously feared that Jupiter would fry its little visitor like a fly in a blast furnace.

Ground personnel in NASA were prepared for the worst. For one thing, they said, Pioneer X was a pathfinder. If it got into trouble, Pioneer XI, which had been launched on April 5, 1973, and was following behind, could be rerouted around the danger area, perhaps settling for something less than a close encounter. For another, NASA had stated in public that Pioneer X really had three jobs: exploring deep space, passing through the asteroid belt, surveying Jupiter. The first two had already been successfully accomplished, the argument went, so that even if Jupiter knocked out the spacecraft, the mission could not be considered a failure. The place and method of Pioneer's death would actually yield valuable information about Jupiter. That is what they said in public.

But in private NASA knew better. They badly wanted Pioneer X to pass by Jupiter alive and safe. They wanted this because they knew the public would consider anything less a failure. Pioneers X and XI cost a total of $100 million! NASA had begun to run into stiff opposition in Congress to the spending of vast sums on work that seemed to many to have no useful benefits to the common man (which is really not true). NASA did not want to have to explain why one of those spacecraft, valued at $50 million, went silent just when things were getting exciting.

And finally, NASA badly wanted to succeed. Ten years of work and genius had gone into the Jupiter missions, and the potential benefits to science could not be overestimated. The people who were beginning to gather near the old blimp hangar at Moffett Field wanted success. They had worked hard and they believed that they deserved it. If hard work, imagination, and considerable care rate at all, they would get it.

So Pioneer X swept through silent, black space. Ahead of it the giant planet and its potentially lethal cloak loomed.

In the middle of November, Pioneer's instruments were turned on and tested. On November 25, they were turned to full power and the 24-hour surveillance began. It was just in time, because Jupiter quickly issued its first surprise.

While Pioneer was still 7,707,000 km from Jupiter, the spacecraft ran into the first obstacle, the bow shock wave, the place of turbulence created when the stream of particles from the sun met the magnetic field of the planet. The magnetic field acted like a wing flying through the air, creating ripples and currents of roughness.

The arrival at the shock wave was duly recorded by the magnetometer, which sent readings three times higher than it had been sending before. Other instruments re-

Control room and the mission control board from which Pioneer is operated

corded that the temperature of the solar wind had jumped by 100 times, which meant that the cold wind was agitated enough by the turbulence to grow very warm. The speed of the solar wind was also affected, braking by half to about 241 km/sec. Soon six of the eleven on-board instruments were sending back data confirming that Pioneer was now approaching the magnetic field itself much earlier than had been expected. The bow shock wave proved to be more than 95,000 kilometers thick.

That same day Ames scientists sent Pioneer a vital message: turn around to the right position to point the instruments at the planet. Like all messages at this stage of the mission, this one was accompanied by great tension among

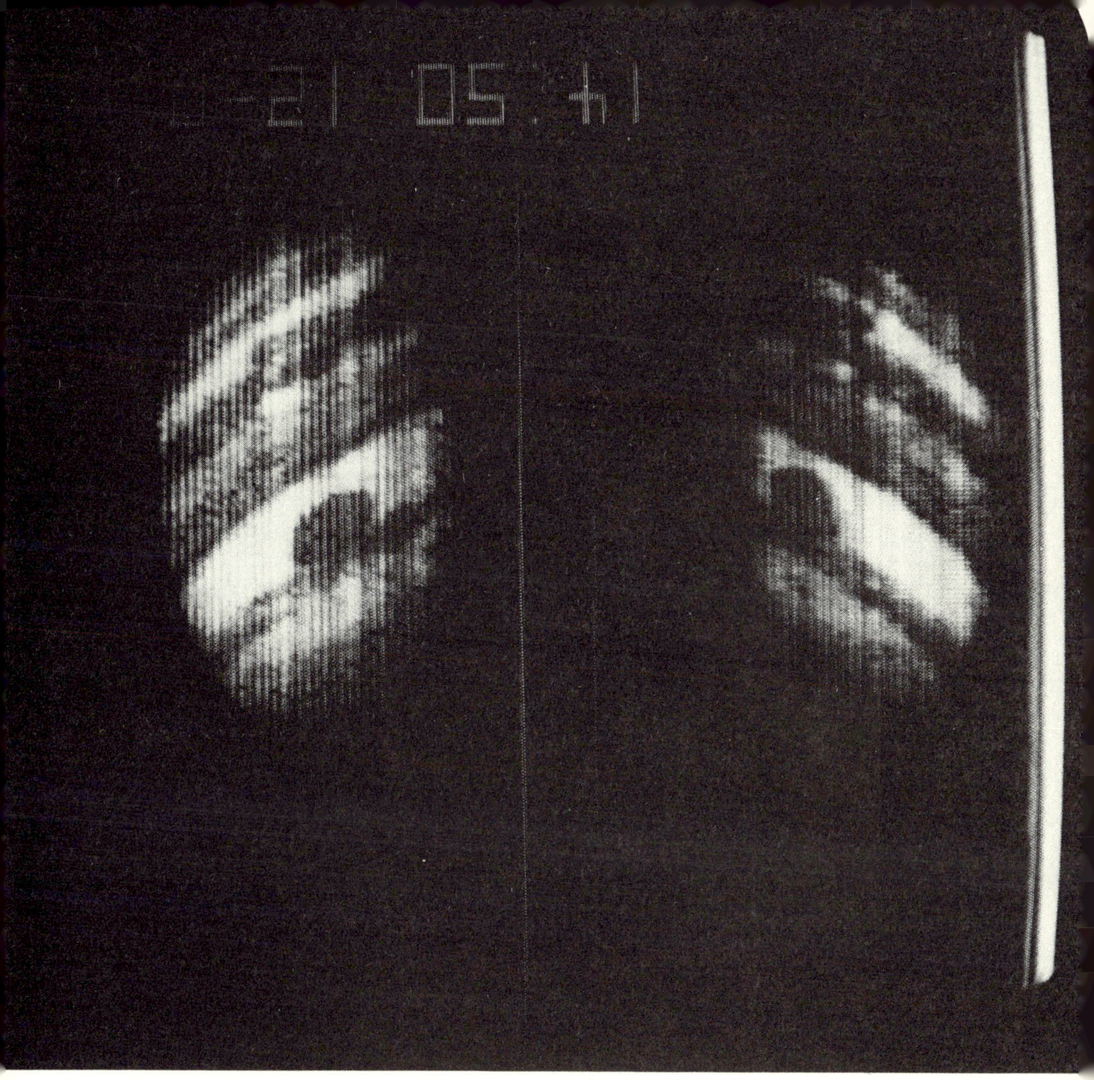

Raw pictures begin to appear upside down on television screens in California (5:51 P.M. PST, 12/1/73)

the operators. Because of the distance from Earth to the spaceship, and even though radio waves travel at the speed of light, it took more than 40 minutes for the radioed command signal to reach Pioneer and another 40+ minutes for confirmation to return. In other words, if anything went wrong during this part of the flight, NASA scientists wouldn't know it until more than 40 minutes later, much too late to do anything much about it. Even if Pioneer blew up, it would be nearly an hour before anyone knew it.

But this message was obeyed perfectly. The spacecraft rotated into the proper position. The confirmation returned on time. All around the mission control center there was a sigh of relief.

The IPP was turned on to full operation, and the incredible pictures began to flow in at a constant rate.

Pioneer had already crossed the orbits of most of the Jovian satellites: Sinope, Pasiphae, Carme, Ananke, Lysithea, Elasa, and Himalia. It had measured Ganymede and Callisto, and the tiny inner moon, Amalthea.

The spacecraft was now 804.5 million km from Earth.

The scientists had been surprised at the sudden appearance of the bow shock wave. They had assumed it would not occur until later in the flight, closer to Jupiter. Now they were in for a second surprise. On Tuesday, November 27, Pioneer smacked into the outer edge of the magnetic field, the magnetopause. The magnetometer showed a large jump and then held steady. The spacecraft was now cruising through the relatively calm field. There was still no sign of damage to the spacecraft or to its instruments. Scientists had expected the magnetopause to appear closer to the planet and farther away from the bow shock wave. Clearly there were going to be more surprises.

Pioneer had now shown that the magnetic field was much larger than anyone had anticipated. This aroused some fears. If the magnetic field had been underestimated, was it possible that scientists had also underestimated the radiation? Was Pioneer in more danger than had been feared? The scientists could only sit and watch the readings pouring in from Jupiter.

The little spacecraft was on its own at the mercy of the huge world.

7
Jupiter Encounter

In the government-modern buildings at Moffett Field, one of the most important scientific events in history was being played out late in 1973. Nestled between the Santa Cruz Mountains and San Francisco Bay, the field and NASA's Ames Research Center facility had become the base station for the exploration of the planet Jupiter. It took a while for everyone at Ames to get used to that.

They walked around talking about something that was happening hundreds of million of miles away to an object that for all practical purposes was on its own.

The titanic planet rolled along as it has since the beginning of time. All that was different was that it now had a visitor.

Most of the world's curious were now watching what happened at Ames and at Jupiter. About one thousand journalists gathered to relay the information to them, to interpret it, to catch the drama of the week's events. Watch now as one of those reporters records the events.

MOUNTAIN VIEW, Calif., Nov. 29—Tiny Pioneer X zoomed through the strong magnetic fields surrounding Jupiter today, sending back breathtaking color pictures of the giant planet.

The flying automatic physics lab is now six days from Jupiter Encounter, man's first closeup look at one of the outer planets of the solar system. . . .

Pioneer project manager, Charles Hall, uses the model to make his point to journalists

The 570-pound satellite has already passed through two zones some scientists thought would endanger the Jupiter mission. . . .

A third place of potential danger, the radiation belts, will not be encountered until five hours before periapsis, the place where Pioneer gets to its closest point to Jupiter, 81,000 miles above the Jovian clouds.

How strong the radiation in the belts will be is unknown and finding out is one of Pioneer's tasks.

A device called the imaging photopolarimeter is sending back full-color pictures of the planet of surprising quality.

The planet appears as a disk against a pitch-black background. It has bands in the clouds, running at a slight angle from left up to right.

The bands are pink, blue, and white on the pictures.

These pictures are raw, unprocessed photos shown here fifteen minutes after they arrive from Jupiter. Later the pictures will be technically augmented by computers at the University of Arizona, which will make them even clearer.

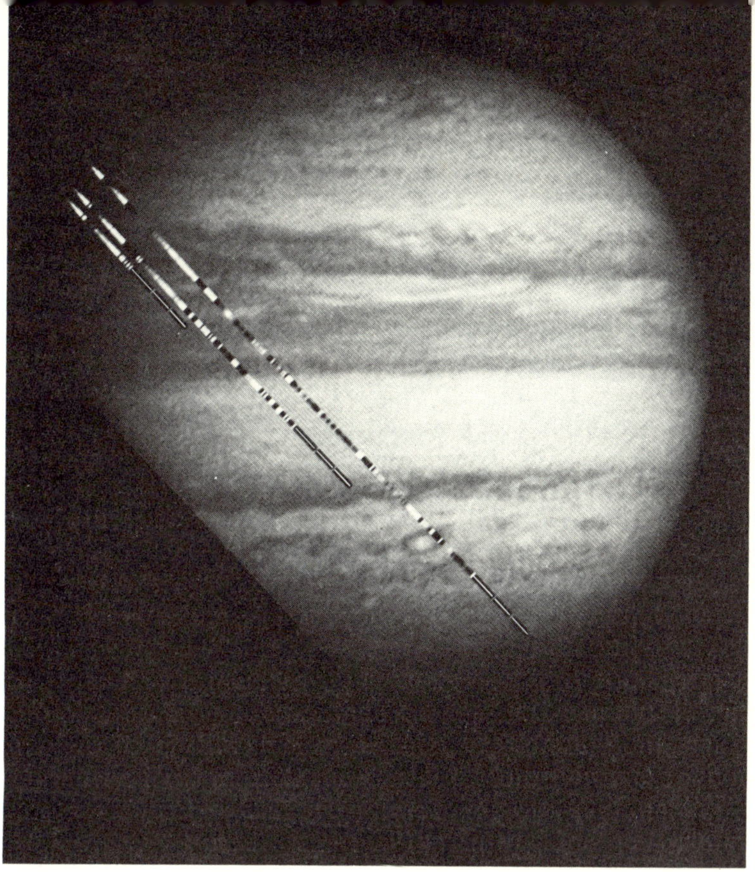

Closer still. The strange rod-shaped lines on the picture result from a loss of data from the computer. Pioneer is now 1.6 million km from Jupiter (12/3/73)

The only distortion that is visible so far is vertical black bands that do nothing to disturb the beauty of the pictures. . . .

Pioneer was now 813,620,000 km from Earth, and 4,-800,000 km. from Jupiter. Its speed, relative to the sun, was 10 km/sec. Radio transmissions took 45.2 minutes.

MOUNTAIN VIEW, Calif., Nov. 30—Pioneer X moved into the fringes of Jupiter's potentially hazardous radiation belts today, having already probed deep into the giant planet's magnetic field.

The 570-pound unmanned physics satellite was three days from Jupiter Encounter, the rendezvous to within 81,000 miles of the Jovian cloud tops.

Color images from the electronic camera on Pioneer

showed a planet growing steadily on the screens at Ames Research Center. The Jovian clouds were bands of pink, light blue, and white with the mysterious Red Spot clearly visible.

All the equipment on the space probe was working perfectly after a year and a half in space. Instrument readings, relayed through computers in Southern California, were being processed here, on the peninsula south of San Francisco.

Dr. James A. Van Allen of the University of Iowa, who discovered the radiation belts around Earth in 1958, which are called the Van Allen belts, said Pioneer now appears to be in an area containing radiation trapped by the magnetic field of Jupiter.

That would mean it is on the fringes of the radiation belts some scientists fear could destroy the complex electronic equipment aboard Pioneer.

The intensity and nature of these belts is almost completely unknown.

Van Allen said the evidence appears to be that the magnetic field of Jupiter is not nearly so round as that of the Earth. It is misshapen on each side.

Since the field is at an angle to the planet's equator, the spacecraft, which is moving on a straight line, is shifting in and out of the stretched ends.

The field itself is something of a mystery.

The Earth's magnetic field is dipolar, which means it has a positive and a negative pole. The northern part is the positive end. It runs at an 11-degree angle from the equator.

According to Dr. John A. Simpson of the University of Chicago, evidence from Pioneer indicates that the magnetic field of Jupiter is also tilted off the perpendicular by about ten degrees. It seems to be a dipole, though he said some evidence from Pioneer is contradictory on this subject.

So far Pioneer has not reached that part of the field which rotates as the planet spins. Scientists said that there doesn't appear to be any increase in the magnetism as Pioneer moves toward Jupiter.

Scientists are finding an increase in electron activity

(negatively charged particles of atoms), but not proton activity (positively charged particles).

That is different from what a satellite approaching Earth would find, scientists said. They had no explanation.

Because of a slight increase discovered in the velocity of the spacecraft, scientists have decided that Jupiter weighs just a little bit more than they thought. The difference is only about 25/10,000th percent of the mass of the planet, well within the limits of astronomers' earlier guesses.

MOUNTAIN VIEW, Calif., Dec. 1—America's Jupiter explorer, which has already driven astronomers here into total confusion with its mounds of data, threw them another shocker today when the space probe reported that the giant planet's magnetic field seemed to have collapsed.

"Oh my God! Don't tell me!" said one scientist when NASA's Dr. John H. Wolfe, a plasma physics specialist, telephoned the latest findings from Pioneer X.

With scientific theories rising and falling in a matter of hours, astronomers had still another mystery to unravel. They were admittedly confused, but obviously having the time of their lives.

Pioneer was giving scientists the greatest quantum jump in knowledge about the solar system's largest planet since Galileo first looked at it in 1609.

This latest discovery occurred last night when Pioneer, which had been sailing through Jupiter's magnetic field since Tuesday, reported that the field had disappeared. It was more than eleven hours before the instruments began reading the field again.

Wolfe said it appeared that the field had collapsed under pressure from a gust of the solar wind—tiny charged particles emitted from the sun.

The outer edge of the magnetic field, called the magnetopause, is kept in equilibrium by the pressure of the solar wind from the outside and from particles trapped on the inside.

Wolfe said a gust of the wind apparently unbalanced this equilibrium and the magnetopause collapsed so fast it passed Pioneer and the tiny space probe was out of the field for eleven hours.

Wolfe said this happens to Earth's magnetic field frequently, but was not expected to happen to Jupiter's just at the precise time Pioneer was flying by.

All systems on Pioneer were working perfectly. The color images sent back through space showed a planet banded in blue, pink, and white with the mysterious Red Spot bright and clear. Even those images which are not being enhanced by computers are awesome, already exceeding what earth-based telescopes can show.

The pictures have also included three of the largest of Jupiter's twelve known moons.

The best example of the confusion from the data concerns the theories about the strength of the magnetic field on Jupiter's surface (if it has one).

Radio astronomers, operating at a huge distance, theorize that the surface magnetism is 7/10 gauss. (A *gauss* is the standard unit of measuring magnetism. The magnetism of Earth's field is ½ gauss.)

Yesterday, based on preliminary data, scientists here said the radio astronomers' figure was probably too low. The correct reading would be as much as 20 gauss.

Within hours they admitted they were wrong. One gauss was going to be closer to the truth.

Pioneer has already established several things astronomers did not know or were not sure of about Jupiter.

The shock wave produced when particles from the sun pile up against the planet's magnetic field is out farther than they thought. The field itself begins farther out than they thought possible.

The field is not like Earth's. It is not rounded but stretched at the ends; it wobbles, and it may be much weaker than thought.

The magnetic field probably tilts like Earth's, but if you stood on the surface of Jupiter (if, indeed, it even has one) and took out a compass, the needle would point south, not north.

As Pioneer slides toward periapsis, readings show radiation trapped within the magnetic field.

Some scientists have felt that the radiation would endanger the spacecraft, but it was too early to tell from readings just how strong they would be.

The radiation varied in many cases every ten hours.

Jupiter rotates once every ten hours, and that is probably the cause of the changes.

But the magnetic field does not appear to be changing with the rotation, which is strange, scientists said.

"The field is probably attached to the planet," said Dr. Sherman DeForrest of the University of California at San Diego, "but it's not rotating with the planet."

He said that means the field is weaker than thought, or it is larger than thought, or none of the above is correct.

Mission conference at Ames Research Center. The participants are, from left, Joseph E. Lepetich, Bernard J. O'Brien, Charles Hall, Ralph W. Holtzclaw, and George J. Northwang, all NASA project managers

By the end of that day, Pioneer was 819,970,000 km from Earth, 2,520,000 km from Jupiter. It was speeding along at 11 km/sec relative to the sun, closing the distance with Jupiter at 13 km/sec. Radio signals were taking 45.6 minutes to make a one-way trip.

MOUNTAIN VIEW, Calif., Dec. 2—Pioneer X moved in toward the most perilous part of its exploration today—Jupiter's probably violent radiation belts—with only one more day until Jupiter Encounter.

The Encounter, scheduled for 9:25 P.M. EST tomorrow night, will bring the tiny space probe to within 81,000 miles of the planet's multihued cloud tops, before zooming into interstellar space.

Pioneer was traveling at 506,800 km/hr relative to Jupiter at 7:00 P.M. tonight. At that time it was 831,139,000 km from Earth and 1.609 million km from Jupiter. It was taking more than 45 minutes for one-way communications because of the distance.

When Pioneer will hit the radiation belts is the great unknown of the mission. Many scientists feared that the intense radiation would zap the unmanned probe, destroying the electronic instruments and communications.

The scientists are running an office pool to bet on when they will see their instruments jump, signifying Pioneer has struck the belts. One physicist has already won two apples and two bottles of gin.

The electron count has been going up for days, but only in the last few hours have the protons increased significantly. It is the protons, which are much heavier than electrons, that pose the greatest danger to Pioneer.

Protons are positively charged particles. Electrons are negatively charged particles. The belts containing these particles are believed to be trapped in the planet's magnetic field.

Only Jupiter and Earth have such fields and, it is felt, have strong radiation belts.

The strength of the fields is continuing to grow, as shown in the instruments. There is still no solid theory as to their nature.

The predominant theory is that Jupiter's field is shaped something like that of Earth, except upside down (north is south). It expands and contracts as the solar wind gusts and blows.

Another theory, which is sometimes called the "flapping pancake" theory, is that the field is fairly flat but set off-center, so that it flaps or wobbles as Jupiter rotates.

Another mystery has to do with the cycles of the particles. The electrons fluctuate in ten-hour cycles, in line with the ten-hour rotation period of Jupiter.

But the protons fluctuate in twenty-hour cycles, twice the rotation of Jupiter, and astronomers here call that "inexplicable."

The answer to all this is important, because Jupiter is a different kind of planet from Earth or Mars. Jupiter appears to be a contracting mass of gas in a stage of creation similar to that which all planets and stars go through at birth.

The average life-span of a scientific theory here is now just a few minutes. Every time one scientist thinks he has an answer, another comes along with a new finding to shoot that theory down.

At 3:00 P.M. EST, Pioneer passed the orbit of the Jovian moon Callisto. At 7:00 A.M. tomorrow it will pass Ganymede, at 1:00 P.M. Europa, and at 5:26 P.M., the moon Io.

Time for communications was now up to 45.7 minutes as the day before encounter ended.

MOUNTAIN VIEW, Calif., Dec. 3—Pioneer X, the tiny spacecraft that may be probing the process of creation in the universe, swung in toward its appointment with the giant planet Jupiter today.

"It's a remarkable moment in the history of mankind," said Dr. Carl Sagan of Cornell University.

The Pioneer space probe is to sail to within 81,000 miles of the roaring tempestuous atmosphere of the planet at 9:25 P.M. EST. Then it will swing out, using Jupiter's immense gravitational field to propel it, toward the outer fringes of the solar system and beyond—the first man-

Nearing encounter. This picture, taken in blue light, shows the cloud tops of Jupiter. The bands come through quite clearly (12/2/73)

made object to reach interstellar space.

By afternoon Pioneer had still not reached the area of the most intense radiation that some scientists feared would destroy the electronic equipment on board, silencing the little explorer. One instrument was changing its operation without commands, but technicians were not sure whether the steady buildup of radiation had anything to do with it.

Pioneer was sending back pictures of the planet far beyond anything possible from Earth. They filled television screens here with the multihued bands of Jovian clouds.

The images so far include two of the huge moons of Jupiter, Ganymede and Callisto. Other instruments have taken Ganymede's temperature.

Scientists here were elated and thoroughly confused by the mountains of data. It is almost impossible to find two scientists who agree on one theoretical model of what the radiation and magnetic fields of Jupiter are really like. Theories come and go in minutes.

The findings are potentially historic.

If, as scientists believe, the Jovian planets are going through the process by which all worlds and all suns are created, the composition of Jupiter may be the original stuff of creation as it was five billion years ago. Additionally, 97 percent of all the mass of the solar system is in the Jovian planets. To understand those planets is to understand most of what the solar system is about.

Pioneer has already measured the hydrogen around Jupiter and has begun to get readings of helium, believed to be the major constituents of stars.

The major scientific controversy so far has sprung up around the magnetic fields, with two theories battling for prominence.

One theory, proposed and defended by Dr. John Wolfe, chief project scientist, has a magnetic field somewhat like Earth's but with its ends stretched. The best way to visualize this is to think of a marble inside an inflated balloon. If you swing the balloon around, the marble distends the rubber, flattening it at the ends.

The second theory, defended by Dr. James Van Allen, is called the "flapping pancake."

Van Allen says the magnetic fields are not spherical at all but in the shape of a disk. It is off-center and wobbles a great deal, like a hula hoop.

The battle of the two theories has been going on for days.

"There is no disk, period," Wolfe said in an interview. "That is completely wrong. The pancake theory is not consistent with the magnetic field and plasma (gas) data.

"I can construct a very consistent picture and I don't need a silly pancake to do it."

But the crackling, beeping sound of Pioneer has already told a great deal about Jupiter and environs.

For instance, it now appears that the magnetic field is slanted by about fifteen degrees and its center is not the center of the planet as is Earth's. That probably says

Flyby. The planet passes below the spacecraft

something about the inner core of Jupiter but no one knows what.

The surface temperature of Ganymede has been measured at −148.3° C.

The intense radiation did not begin until 2.7 million km out.

Encounter!

MOUNTAIN VIEW, Calif., Dec. 3—Pioneer X, braving the strong radiation belts around Jupiter, sailed to within 81,000 miles above the roaring atmosphere today, probing the processes of creation.

The radiation around the tiny explorer peaked about ten minutes before periapsis, and then receded, to the general surprise of astronomers here. Many had feared the radiation would silence the little space probe.

After rendezvous, Pioneer began to sail away from the giant planet, heading out toward interstellar space. It had opened what may be a golden age of planetary exploration. . . .

Success!

MOUNTAIN VIEW, Calif., Dec. 3—Pioneer X, braving the roaring radiation belts of Jupiter, plunged to within 81,000 miles of the planet's cloud tops, and then sailed out toward interstellar space.

All the while the 570-pound satellite sent back mountains of data and closeup pictures of our sun's biggest planet.

"We're still alive," said Dr. John Wolfe, Pioneer project scientist, as the radiation readings began to level off. Many scientists had feared that the radiation would cripple the solid-state circuitry on board the little explorer, but this didn't happen.

Pioneer reached its closest point to Jupiter at 9:46 P.M. EST, but it was not until 46 minutes later that readings from that point came in.

The radiation was in belts circling the planet. They consist of electronically charged particles of atoms trapped by Jupiter's magnetic field flowing in currents.

"Needless to say, we're delighted," said project manager Charles Hall.

Wolfe said that after periapsis there had been no damage to the spacecraft from the intense radiation. He warned that Pioneer still had to get away from Jupiter unharmed but pointed out the worst was over.

The place of highest radiation was along a plane from the equator, and Pioneer would not pass that point again. Wolfe said that the readings just before periapsis were 10 million high-energy electrons per square centimeter per second. Readings of the more dangerous protons were 40,000 per square centimeter per second.

"The effects are not as detrimental as perhaps we expected," Wolfe said.

"The Earth's radiation belts are really quite mild com-

The man who ran the show, Charles Hall, Pioneer project manager

pared to what we went through," he added.

Wolfe contradicted several conclusions reported earlier. The latest readings indicate that the magnetic field tilts by ten or eleven degrees, just like Earth's, not the fifteen degrees reported earlier. It was also incorrect to say that the center of the field is not in the center of the planet, he said.

These contradictions have been typical of what has been happening here all week as astronomers try to cope with a torrent of data.

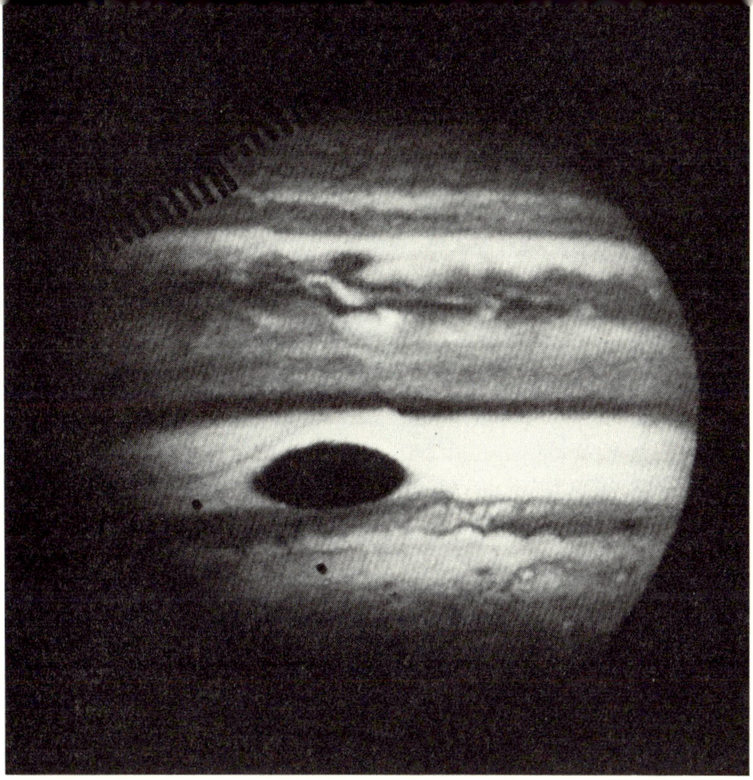

The Great Red Spot and the flow of atmosphere around it in the South Tropical Zone (12/2/73)

Pioneer took pictures of three of the four major (Galilean) moons of Jupiter. It also took pictures of the mysterious Red Spot, a huge object that floats on the southern hemisphere. That picture was made as the spot was in the area of sunset so that a difference in height could be spotted. . . .

Only the next day did scientists realize how close they had come to failure.

MOUNTAIN VIEW, Calif., Dec. 4—NASA scientists said today that Pioneer X came frighteningly close to being destroyed by roaring radiation belts of Jupiter.

"We could hardly have played the game any closer," said Robert S. Kraemer, director of planetary programs for the space agency. He added, "The mission, by all standards, was 100 percent successful."

The 570-pound unmanned explorer, using the gravitational field of Jupiter as a slingshot, was sailing toward the edge of the solar system. At 7:00 P.M. last night it was 1.4 million km from Jupiter, leaving the planet's environs at 57,300 km/hr.

Pioneer left behind mountains of data that will take scientists years to unravel. It also produced superior pictures of the planet.

At the peak of the radiation burst scientists were recording 500,000 *rads* (radiation units) on unprotected instruments aboard Pioneer. Five hundred rads would kill a man, so no human could have survived the flight of Pioneer X.

What apparently saved the spacecraft, according to Dr. R. Walker Fillius of the University of California at San Diego, was the fact that the most intense parts of the belt were in a line with the magnetic equator of Jupiter. Because the magnetic field is tilted, Pioneer came quickly up through the plane of the equator about an hour before its closest approach to Jupiter, so it never reached what might have been even higher radiation counts.

These findings would not make manned flights to Jupiter impossible, scientists pointed out. Instruments on Pioneer that were shielded had one-fifth the rads of the unshielded instruments. It is also possible to avoid the equatorial plane during a future flight, thereby missing the worst of the radiation.

"We were biting our nails to see if the spacecraft would survive," Fillius said, "and it did, by a hair. The trajectory was just as close in as we possibly could have gone and still survive."

Other findings were still awaiting further interpretation:

—Pictures of the Great Red Spot appear to support the idea that it is a towering pillar of cloud rising several kilometers above the upper cloud deck.

—The temperature at the cloud tops was −133° C. The temperatures of the moons Ganymede and Callisto were even lower, −145° C for Ganymede, −163° C for Callisto.

—There was some variation in the temperatures within Jupiter's clouds which may eventually help tell something about the interior of the planet.

—The magnetic field is weaker than suspected. Scientists are still unable to agree on its shape but agree that it is upside-down, larger than expected, and slightly tilted from the perpendicular.

—Pioneer lost four pictures due to mechanical problems, including the best possible picture of the moon Io. There were unexplained problems with the camera that occurred both going toward and away from Jupiter.

Scientists at Ames checking and photographing the images as they come in from Jupiter

Pioneer X sailed on. The scientists went back to their laboratories with crates of new data. One year later the scientists would return to watch Pioneer XI sail by the planet on its way toward Saturn. More crates of information would have to be carted back to laboratories and computers.

Pioneer did the job.

8
Jupiter

It did take the scientists years to make some sense out of the data returned by the two Pioneer spacecraft. To a large extent they have not finished yet. But by combining the results of the Pioneer missions with the best of earth-based astronomy, we now can come to a fairly complete picture of Jupiter and of the structure of the Jovian planets. Further exploration, and perhaps someday a visit by astronauts, will be needed before anyone can really be sure, but until then we have this picture of the sun's giant.

JOVIAN WEATHER

The violent, almost brutal weather in the Jovian clouds fascinates scientists. In many ways it is drastically different from Earth's weather. In other ways, however, it can provide a clue to understanding what happens in our own skies.

Unlike the weather on Earth the weather on Jupiter either flows around the equatorial and temperate zones or stays localized in the polar regions. On Earth, weather flows from the equator toward both poles and back again.

By studying Jovian weather, however, the scientists can watch the Earth's tropical regions on a large, very unsubtle, scale.

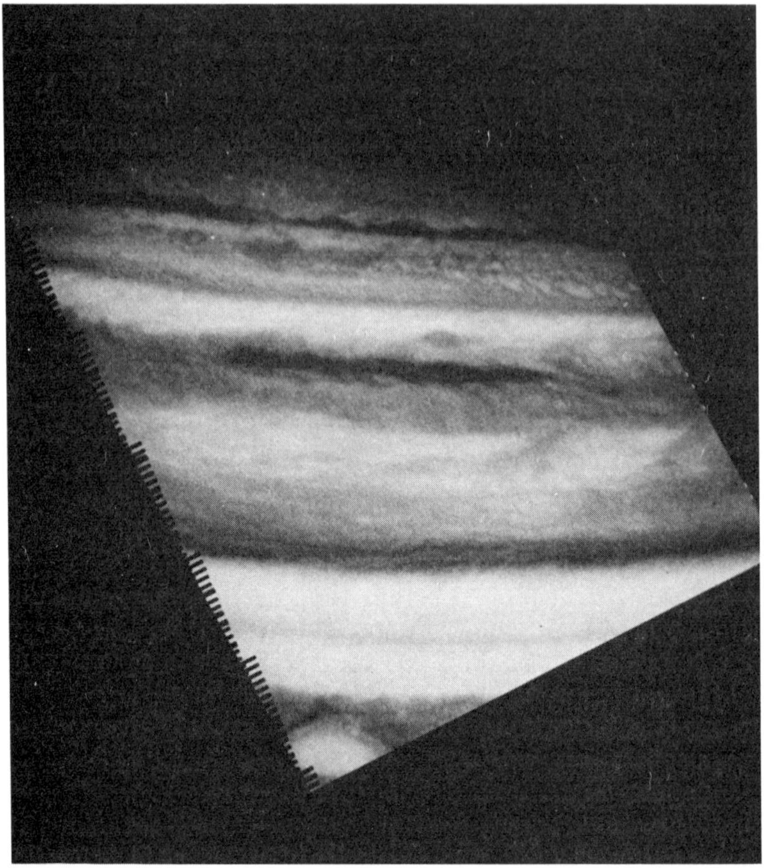

Closeup of Jupiter's Equatorial Zone. Streaks of streaming clouds mark the zones and belts, which appear somewhat fuzzy in this picture. Many of the features are hundreds of miles long. Pioneer is 814,000 km from the planet (12/3/73)

According to data from Pioneer XI, Jupiter also lacks the temperature variations found on Earth. The atmosphere is heated uniformly from pole to pole. On Earth it is much warmer at the equator than at the poles.

What makes Jovian weather so fascinating is that you can see it all. Unlike the Earth, where most of the weather is made up of invisible clear air, Jupiter's atmosphere is all clouds. That means weather patterns, currents, twists, and bends are easy to see. The bands and belts are patterns of flow within the clouds, patterns that last twenty or thirty years or more.

The polar regions of Jupiter behave the way equatorial regions do on Earth. The polar formations are caused by warmer gases rising, cooler gases falling. It is a tumultuous area, filled with relatively small features—more than a thousand miles across—that are the Jovian equivalent of cyclones. Some of the heat driving these convection currents may come from the internal heat of the planet warming the gases.

What drives the weather system planet-wide is still not known, but in the equatorial and temperate zones, the areas between 50° North and South latitude, it appears to be convection again. The convection is contained in a relatively small part of the atmosphere, a layer that contains three large cloud layers. The bottom layer is composed of regular water clouds, just like Earth's. Below that layer are uniformly heated zones of compressed hydrogen and helium gas. Below that is the liquid core of the planet.

The main flow of the upper atmosphere weather is horizontal—east and west. This is caused by something called the *Coriolis forces.* Coriolis forces occur when a gas or a liquid is spun around a globe or in any circular fashion. They cause water going down a drain in a sink to rotate clockwise in the northern hemisphere of the Earth, counterclockwise in the southern. On Jupiter, they move the outer atmosphere.

Inside all of this, *cells of convection,* which are masses of rising heated gas or falling cooling gas, form and spread out. Eventually they reach all around the globe and the ends merge. That process forms the zones and belts of Jupiter.

As with Earth, it is heat that drives the Jovian weather system. Pioneer XI found that 40 percent of the heat used to drive the weather engine of Jupiter is internally produced—it comes from within the planet. The other 60 percent is generally absorbed from sunlight, even at Jupiter's great distance from the sun.

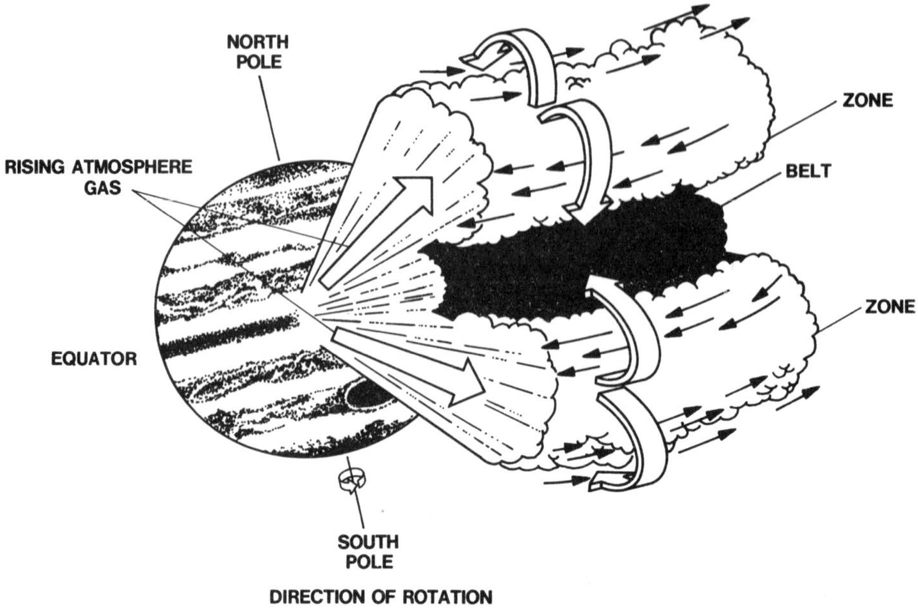

How the weather system works on Jupiter. Convection currents, rising warm gas, causes the atmosphere to roil the way bubbles in boiling water cause it to stir. The zones are the high points, the belts are the valleys in between. The wind passes in each direction

The polar regions get more of the internal heat because they are not in the direct line of sunshine. Since Jupiter is mostly a liquid planet, and liquids tend to even out temperatures, most of the internally generated heat flows toward the poles.

The winds howl on Jupiter. Telescope measurements over the last twenty years indicate that winds in Jupiter's zones roar along at 320 km/hr and can reach 600 km/hr. The winds blow alternately eastward and westward in the zones, which are generally considered areas of high pressure, warmer temperatures, rising gases, and thicker cloudiness. The belts, on the other hand, are regions of lower pressure, cooler temperatures, falling gases, and thinner clouds. According to Pioneer, the top of the clouds in the zones is higher than in the belts.

There are other explanations for the driving force of Jupiter's atmosphere. One is that the planet is so large that when it spins rapidly the atmosphere naturally breaks

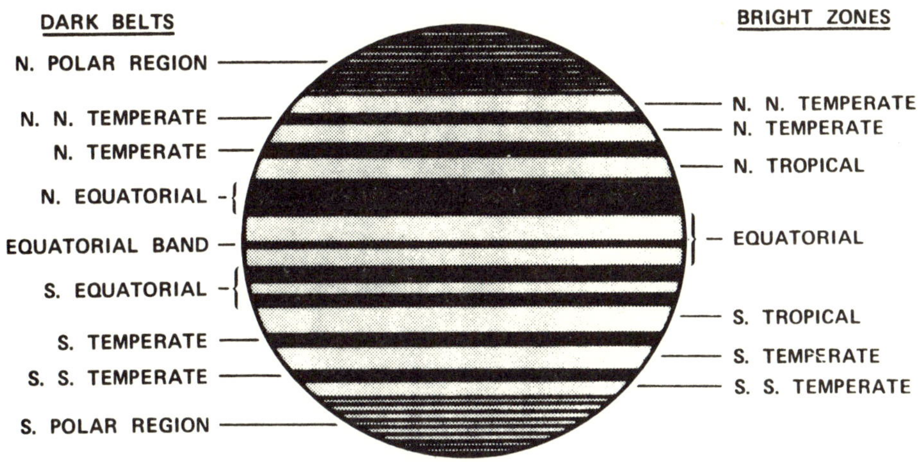

Belts and zones

down into zones and belts. Very likely not only one "engine" is at work here, but a combination of many forces. It will be a long time before the mystery is unraveled to everyone's satisfaction.

Here are some other findings about Jupiter's atmosphere:

—Temperatures decrease by an average of 2° C for every kilometer of altitude.

—The content of the planet in carbon, nitrogen, and helium is the same proportionately as the content of the sun, confirming the theory that the solar system was all created from the same cloud.

—Because of the lack of friction or physical obstacles on the "ground," structures in the weather pattern tend to stay for long periods of time. One eddy or current could last for centuries without major alteration.

—The atmosphere is about 70 kilometers thick. The lower part of the atmosphere is considerably quieter than the upper parts because there is less disturbance there to stir it up.

—Atmospheric pressure within the three-layer skin of

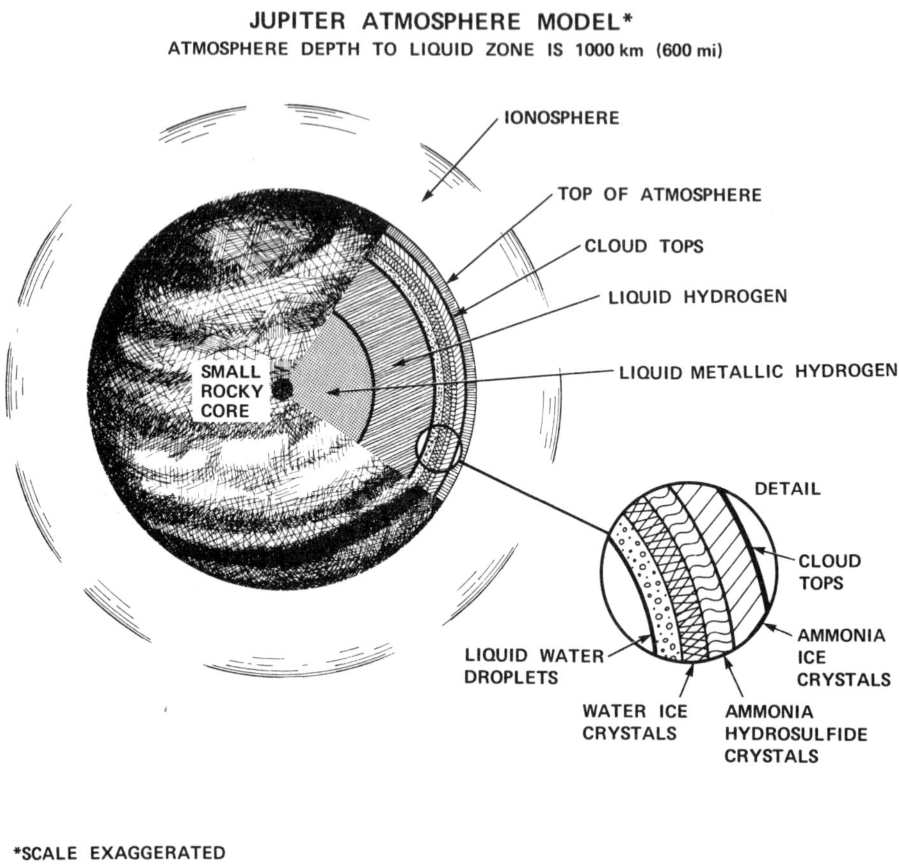

The atmosphere, showing the core

the atmosphere ranges from ½ of Earth's pressure at the top of the clouds, to 4½ at the bottom.

—The top of the atmosphere is a layer of ammonia crystals and ammonia gas. Below that is a layer of ammonia hydrosulfide crystals. At the bottom are water ice crystals and water vapor.

—The reddest clouds are the highest, they are red because they have picked up phosphorous. White clouds are the next highest, getting their color from ammonia. The gray or brown clouds come next in altitude, getting their coloration from complex organic chemicals and from sulfur, hydrogen, and ammonia.

—Under the atmosphere are compressed hydrogen and helium at a ratio of 88 to 12.

—The visible cloud tops are made of ammonia ice crystals at an atmospheric pressure of 7/10 of Earth's and a temperature of −120° C.

—There is a "blue sky" near the polar regions, the transparent stratosphere is detectable.

—Fourteen km above the cloud tops the pressure is 3/10 of Earth's atmospheric pressure and the temperature is down to −145° C.

THE GREAT RED SPOT

The Red Spot remains something of a mystery, but Pioneer told us enough about it to permit some very educated guesses. The spot appears to be an unusual atmospheric

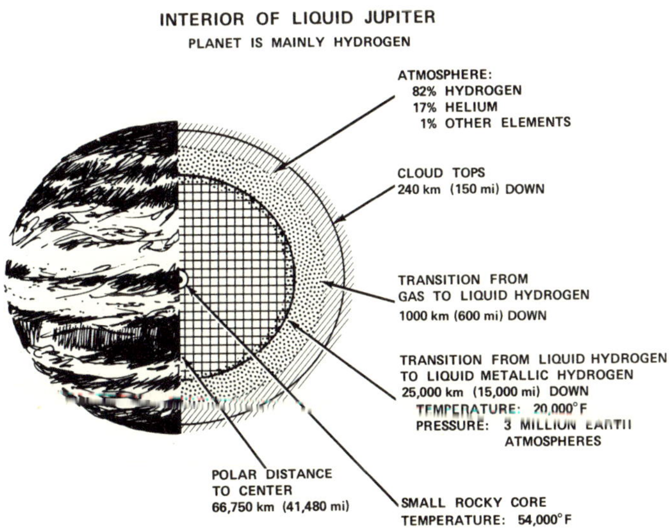

Anatomy of the giant. This is the best guess based on what we learned from Pioneer

feature. For the last three hundred years it has been turning counterclockwise, splitting the northern and southern halves of the South Tropical Zone.

The red cloud that sits on top of the Spot is the highest feature of Jupiter's visible atmosphere, rising about 8 km above the surrounding cloud deck. The red color comes from phosphorous, which apparently condenses at that altitude and temperature.

The idea that the Spot is fixed to some structure in the interior of the planet does not seem to be possible. The Spot slowly moves westward, which it would not do if it were attached to something.

It is probably a column of rising air, a "super zone," in the words of one scientist.

Four much smaller red spots have been discovered in the past few years, giving some support to the "super zone" theory.

MAGNETIC AND RADIATION FIELDS

The great debate over the pancake or the stretched balloon theories of the magnetic field have now been replaced by the "leaky bag" theory.

Think of a balloon filled with water being whirled around. Every once in a while a droplet leaks out and goes flying off into space. That appears to be what happens with Jupiter's magnetic field. Instead of water, however, high-energy electrons go flying off, off as far as Mercury. Because it requires 100 billion watts of energy to shoot these particles away from Jupiter's gravitational field, it is safe to assume that Jupiter is one of the largest sources of high-energy radiation in the solar system.

Where does the energy come from? Apparently, the particles bounce from the inner belts out to the edge of the magnetosphere and back, over and over again. Each time

JUPITER'S MAGNETOSPHERE

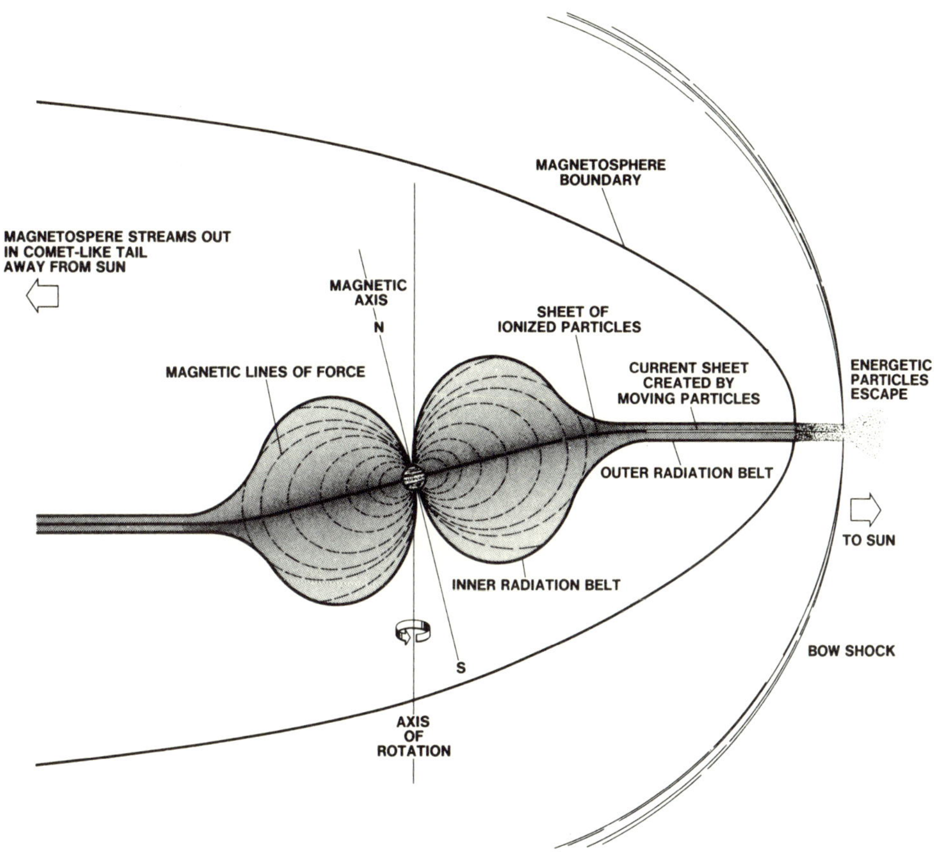

Drawing of Jupiter's magnetosphere

they pick up some more energy until they have enough to break loose and hurtle into space.

The flopping motion of the magnetosphere comes from pressure exerted by the solar wind. It pushes the back end away from the sun, forming something that looks like the tail of a comet.

Many of the particles in the magnetosphere are soaked up by the outer edge of the atmosphere and by Jupiter's moons. If it were not for that and for the escape of so many particles, it is likely that the magnetic fields would be as much as one hundred times as powerful as they are.

The inner layer of the magnetic field is shaped like a doughnut with the planet in the middle, just as with Earth. But it is ten times as powerful as Jupiter's cloud tops. There is one phenomenon at work on Jupiter that is not present on Earth—three of Jupiter's moons act like lightning rods, soaking up much of the radiation. Future astronauts will bless those moons. They sweep up 99 percent of the heavy protons that would be deadly to any living thing that happen by. They would also probably be deadly to electronic equipment in automated spaceships like the Pioneers. Without those moons the Pioneer missions would have failed.

Pioneers X and XI succeeded, however, and here are some of the statistics and images they sent back:

—The strength of the magnetic field is 17 times that of Earth. The total amount of energy, however, is 20,000 times that of Earth because of the difference in planetary size.

—The field is tilted 11 degrees off the axis.

—Oddly, the center of the field is not the center of the planet. The center of the field is actually located north of the center of the planet and slightly outward from it, a total of 7,140 kilometers away.

—It wobbles by as much as 22 degrees because it isn't parallel to Jupiter's equator. The wobble goes through 10-hour cycles.

—The average diameter of the magnetosphere is 14.5 million kilometers.

—The magnetosphere has three distinct regions: the inner region with the most intense radiation, the middle region containing a flat layer of electrified particles like an electrified disk, and an outer layer that goes as far as 614 million kilometers from the planet.

RADIATION BELTS

The inner radiation belts of the planet really sizzle. They are as much as one million times as powerful as the famous Van Allen belts of Earth and as strong as atomic bomb blasts. They are the most intense measurements ever made in space.

Had astronauts been aboard Pioneer X, they would have been fried. In fact, the radiation was almost strong enough to roast the electronic equipment. Because of the angle at which it approached Jupiter, Pioneer X received more radiation than did Pioneer XI, which actually came closer. The dosage of Pioneer X was one thousand times the lethal dose for humans.

Worse than this, three of the four giant moons are within the most dangerous parts of the radiation belt, so landing humans on those moons is going to be very risky.

Pioneer X received six million protons per square centimeter per second. Although Pioneer XI received twenty times as much in intensity, the belts it flew through were quite thin and the spacecraft was in them for only a short period of time. The total radiation was less than that received by Pioneer X.

The highest levels are on a plane with the magnetic equator. Above and below that, the drop-off is quite steep.

IO

It turns out that the moon Io is the Jupiter Broadcasting System after all.

The mystery was the source of the gigantic decametric outbursts that were the equivalent of a couple of hydrogen bombs. Io is the answer.

It is a case of turning a cosmic electric switch. Io sails through space above Jupiter's ionosphere. The moon is electrically conductive and it attracts electricity from Jupiter's ionosphere. When enough energy builds up, Io snaps out a bolt of lightning and the sound of it, in the form of radio waves, can be heard all the way to Earth. If you were standing on Io, the sight of it, in the form of unimaginable bolts of lightning, would be spectacular and terrifying. Think of 10-trillion-volt flashes!

Scientists at the University of Hawaii figured it out. They discovered that the radio blasts occurred every twenty-one hours, which is half the time required for Io's orbit around Jupiter. Io was either on the Earthward side or behind Jupiter at the time.

The conductivity comes from the presence of electrically charged sodium in Io's ionosphere and in the space around it. The sodium had been blasted from Io when radiation particles crashed into the salt plains on the surface.

LIFE

As we said, the possibility of life in Jupiter's vicinity is not so weird as it sounds. It turns out that all the ingredients are there and only part of Jupiter's atmosphere would be deadly to life.

Life would have begun, if it ever did, in the atmosphere between the place where temperatures are cold enough to freeze water and where they are high enough to boil it. The atmospheric pressures at those levels range from five to ten times Earth pressure.

The problem life would have is that the roaring convection currents would eventually send the organisms down into the hottest parts of the atmosphere where they could not possibly survive. To get around this problem one scientist suggested that the microrganisms might simply learn to reproduce fast enough to keep the cycle going despite the loss of some of the population to the heat.

That's not the only problem life would face. The high concentrations of hydrogen in the atmosphere may keep down concentrations of the tiny organic molecules that make up the building materials for life.

What makes many scientists think life could have begun at Jupiter is simply its size. There is so much of it that enormous quantities of the raw materials are being produced, a thousand times the area of Earth's life-birth. There are also vast amounts of energy at Jupiter, another requirement for life.

There are probably pools of water in the atmosphere in which the organisms could blossom.

What would happen is that ultraviolet radiation from the sun would chemically react with elements in the Jovian atmosphere, producing some of the simple organic molecules required. These molecules would probably be red in color, and some scientists think that the red shading in parts of Jupiter is at least partially the result of such molecules.

If enough of these molecules were formed, some might survive circulation into the very hot parts of the atmosphere, provided circulation is very slow.

If circulation is fast, the ultraviolet radiation would not do the job, but visible light from the sun might cause a form of photosynthesis, the same reaction by which Earth plants produce energy from sunlight.

Where would the organisms exist? Riding on particles of dust, probably. Although scientists are talking about the largest planet in the solar system, they are also talking about minute forms of life, if indeed, there is life at all.

MOONS

Since the flights of Pioneer, two more Jovian moons have been discovered, both by Dr. Charles Kowal of the

Jupiter from Pioneer XI. Note the satellite

Hale Observatories on Mt. Palomar in California. Kowal, who specializes in studying *supernovae,* which are giant exploding stars, began studying the area around Jupiter in 1974. He used a special frosted glass mask on his 122-centimeter Schmidt telescope to block out the glare from the giant planet.

On September 10, 11, and 12, he took pictures of the area around Jupiter and on the fourteenth he had them developed and ready for study.

Kowal found a smudge on the prints that he thought was likely to be a satellite, 4.8 to 8 km in diameter. It was, he thought, between Sinope, the twelfth moon out, and

Lysithea, the eighth. He designated it J-XIII. Since then it has been found to be the sixth moon out from Jupiter, between Callisto and Himalia.

Kowal believed that J-XIII could have been part of a larger body that had somehow been shattered. There may be other pieces of it circling Jupiter.

Kowal's discovery was the first since 1966, when the tenth moon of Saturn, Janus, was found. Since his discovery the International Astronomical Union asked Kowal to name J-XIII. He chose Leda, which they accepted, and that is now the satellite's official name.

But Kowal wasn't done with Jupiter. In October 1975 he found another smudge on prints taken September 30, October 1, and 2. It was probably another moon, J-XIV, but if so, it was the dimmest moon in the solar system by far. Other astronomers have since found the object, but it is not yet officially recognized as a moon and has yet to be assigned a name.

Postscript

In the summer and early fall of 1977, America's space agency launched two more unmanned spacecraft toward Jupiter. They are called Voyager.

These craft are expected to arrive at Jupiter in 1979 and continue on to Saturn and possibly even Uranus, taking more pictures and more measurements, refining what we know about these worlds.

Like their predecessors, the Pioneers, they will eventually sail out into the cosmos. In about 40,000 years they may even approach another star.

Like the Pioneers, the Voyagers carry messages to any alien life-forms who may wander by and find them. In this case the messages are contained in a phonograph record (with a needle and instructions for playing). The record is called "Sounds of Earth," and it contains examples of our languages, our music, and the sounds of terrestrial nature. There are also pictures of everything from childbirth to volcanoes. It is a graphic and aural atlas of Earth and life on Earth at the end of the twentieth century. By the time it is found—if it ever is—that life may be dust.

Also contained aboard the Voyagers is a message from the President of one of the biggest countries on Earth at that time, a country called the United States of America. That President expresses the hopes of the people who built the spacecraft as well as those of us who paid for it. He says in part:

"This is a present from a small distant world, a token of our sounds, our science, our images, our music, our thoughts and our feelings. We are attempting to survive our time so we may live into yours. We hope someday, having solved the problems we face, to join a community of galactic civilizations. This record represents our hope and our determination, and our goodwill in a vast and awesome universe."

Acknowledgments

Some debts are created when anyone writes a book. Books are never written by one person alone, and this book was no exception.

The pictures came from NASA–Ames' Audio-Visual Department and a friend, Jack Stewart, who runs it.

Chapters 6 and 7 were reported while I was on assignment for *The Philadelphia Inquirer* in California. The *Inquirer's* National Editor at the time, Steve Seplow, was responsible for sending me there. Parts of Chapter 7 appeared in the *Inquirer* in a slightly different form.

Another friend, George Hamilton, of the Franklin Institute's Fels Planetarium, read the manuscript for accuracy. If any errors slipped through, the responsibility is all mine.

Most of all, of course, I am indebted to the several thousand engineers and scientists who built and operated the remarkable Pioneer spacecraft. Robot exploration is something new in history, and as I hope you will see, it can be just as exciting as the old-fashioned kind. Pioneer was a brilliant piece of work and I am very proud to know many of the people who were responsible for it. We all owe them our gratitude.

Finally, I am grateful to Jonathan and Michael Shurkin, whose curiosity about everything under the sun is enough to keep any father on his toes. Their awe is part of the reason I wrote this book.

Index